Interaction of
Mid-Infrared Parametric
Waves in Laser Plasmas

Interaction of Mid-Infrared Parametric Waves in Laser Plasmas

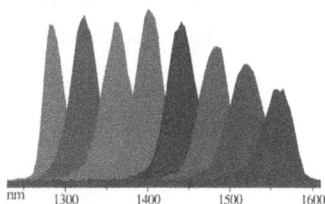

Rashid Ganeev
Voronezh State University, Russia

World Scientific

NEW JERSEY · LONDON · SINGAPORE · BEIJING · SHANGHAI · HONG KONG · TAIPEI · CHENNAI · TOKYO

Published by

World Scientific Publishing Co. Pte. Ltd.

5 Toh Tuck Link, Singapore 596224

USA office: 27 Warren Street, Suite 401-402, Hackensack, NJ 07601

UK office: 57 Shelton Street, Covent Garden, London WC2H 9HE

Library of Congress Cataloging-in-Publication Data

Names: Ganeev, Rashid A., author.

Title: Interaction of mid-infrared parametric waves in laser plasmas /
  by Rashid Ganeev (Voronezh State University, Russia).

Description: Singapore ; Hackensack, NJ : World Scientific, [2017]

Identifiers: LCCN 2016054700| ISBN 9789813208254 (hardcover ; alk. paper) |
  ISBN 9813208252 (hardcover ; alk. paper)

Subjects: LCSH: Laser-plasma interactions. | Laser plasmas. | Laser beams. |
  Nonlinear optics. | Infrared radiation.

Classification: LCC QC718.5.L3 G37 2017 | DDC 530.4/46--dc23

LC record available at https://lccn.loc.gov/2016054700

British Library Cataloguing-in-Publication Data

A catalogue record for this book is available from the British Library.

Preface

The goal of this book is to show the most recent findings of the newly emerged field of high-order harmonic generation using the mid-infrared pulses from parametric amplifiers propagating through the narrow and extended laser-produced plasma plumes. The main difference between this book and monographs on plasma harmonics is related with the application of the long-wavelength source for the high-order harmonic generation in various plasma media, which allowed observation of new peculiarities of the harmonic generation in ablated media.

All recent amendments of plasma harmonic studies could not have been realized without the collaboration between various research groups involved in the studies of the nonlinear optical properties of ablated species. Among numerous colleagues I met and had the privilege to collaborate with I would like to thank H. Kuroda, M. Suzuki, S. Yoneya, M. Baba, F. Mitani (Saitama Medical University, Japan), T. Ozaki, L. B. Elouga Bom, J. Abdul-Haji, F. Vidal (Institut National de la Recherche Scientifique, Canada), P. D. Gupta, P. A. Naik, H. Singhal, J. A. Chakera, R. A. Khan, U. Chakravarty, M. Raghuramaiah, V. Arora, M. Kumar, M. Tayyab (Raja Ramanna Centre for Advanced Technology, India), J. P. Marangos, J. W. G. Tisch, C. Hutchison, T. Witting, F. Frank, T. Siegel, A. Zaïr, Z. Abdelrahman, D. Y. Lei (Imperial College, United Kingdom), M. Castillejo, M. Oujja, M. Sanz, I. López-Quintás, M. Martín (Instituto de Química Física Rocasolano, Spain), H. Zacharias, J. Zheng, M. Wöstmann, H. Witte (Westfäliche Wilhelms-Universität, Germany), T. Usmanov, G. S. Boltaev, I. A. Kulagin, V. I. Redkorechev, V. V. Gorbushin, R. I. Tugushev,

N. K. Satlikov (Institute of Ion-Plasma and Laser Technologies, Uzbekistan), M. B. Danailov (Sincrotrone Trieste, Italy), B. A. Zon, O. V. Ovchinnikov, M. V. Frolov (Voronezh State University, Russia), D. B. Milošević, S. Odžak (University of Sarajevo, Bosnia and Herzegovina), M. Lein, M. Tudorovskaya (Leibniz Universität Hannover, Germany), E. Fiordilino, D. Cricchio, P. P. Corso (Università degli Studi Palermo, Italy), V. Tosa, K. Kovács (National Institute of R&D Isotropic and Molecular Technologies, Romania), V. V. Strelkov, M. A. Khokhlova (General Physics Institute, Russia), M. K. Kodirov, P. V. Redkin (Samarkand State University, Uzbekistan), A. V. Andreev, S. Y. Stremoukhov, O. A. Shoutova, (Moscow State University, Russia), M. Y. Emelin, A. S. Emelina (Institute of Applied Physics, Russia), Pengfei Lan, Zhe Wang, Peixiang Lu (Wuhan Institute of Technology and Huazhong University of Science and Technology, China), A. Husakou (Max Born Institute for Short-Pulse Spectroscopy and Nonlinear Optics, Germany) for their activity in the development of this relatively new field of nonlinear optics.

My family, Lidiya, Timur, Dina, Anya, and Timofey are all what I have in this world. Being a scientific tramp and spending most of my life on various trips I am always with my family in my thoughts and thank them for overwhelming support of this *modus vivendi*. As husband, father, and grandfather, I feel that it is my privilege to serve my family in any capacity.

Rashid A. Ganeev
Voronezh, Russia
January 2017

Contents

Introduction

The application of mid-infrared (MIR) coherent radiation for analysis of various optical and nonlinear optical properties of a large class of materials provides the attractive opportunity to define their properties, which could be useful for various applications in physics, chemistry, biomedicine, and other fields of science, as well as practical use in different branches of industry. Interest in these applications is also based on the need to understand the fundamental principles of laser-matter interaction. Among them, the use of already developed technique of attosecond spectroscopy of gases for implementation in laser-produced plasmas may open new doors for further understanding of the ultrafast processes in the extremely broad community of solids being ablated by laser radiation. Taking into account the prevalence of a few ten thousands of solid materials compared to a few tens of gases, one can anticipate the advantages in the development of laser-produced plasma spectroscopy using nonlinear optical methods.

The solids, being "properly" ablated, transform in the plasma plumes, which could be easily analyzed using the method of high-order harmonic generation (HHG) during propagation of the ultrashort pulses through these plumes. One can define the most suitable conditions for plasma formation for these purposes. This medium should be low-dense (i.e. not exceeding 5×10^{17} cm^{-3} concentration of plasma particles) and weakly ionized (i.e. not exceeding 8×10^{16} cm^{-3} concentration of free electrons). There are a few outcomes from this HHG method: (a) formation of the short wavelength sources of coherent radiation; (b) analysis of the efficiency of this process depending on the structural, morphology, chemical, etc. properties of the solids; (c) definition of the energy structure of the ionic transitions of ablated solids depending on the nonlinear optical response of the media, and investigation of the ionic and atomic transitions possessing strong oscillator strengths through the growth of conversion efficiency of a single harmonic in the vicinity of these transitions; (d) formation of clusters and nanoparticles during laser ablation of solids and analysis of their influence on the strength of the harmonics generated by these species, as well as ablation of micro- and nanoparticles already existing on the surfaces of solid materials for the studies of the spatial characteristics of harmonic emitters; (e) formation of periodically modulated plasma jets for quasi-phase-matching (QPM) of interacting waves in these heterogeneous structures and generation of a few enhanced harmonics in different ranges of extreme ultraviolet (XUV); (f) analysis of interference processes of different quantum paths during acceleration of electrons in laser-produced plasmas; (g) generation of ultrashort (attosecond) pulses during HHG in plasmas and corresponding spectroscopy of plasma plumes using these pulses.

Some of above issues [(a), (f), (g), and to some extent (b)] have already been partially developed using the development of HHG-in-gas concept. In this connection the question "why do we need HHG-in-plasma?" may arise provided the proposed concept does not offer new knowledge in material science. This question is closely related with further directions in one of the branches of nonlinear optics. Indeed, why plasma harmonics? Do we need plasma harmonics if we have a well-elaborated technique of frequency conversion of laser radiation towards a short-wavelength range through nonlinear optical HHG in gaseous media? One can recall that there is also another technique of HHG, i.e., surface harmonic generation. Both of them have a relatively long history of developments and applications for various needs and seem to fully satisfy the scientific community as sources of coherent XUV radiation. Let us consider the problem met by two abovementioned conventional HHG approaches and discuss what the third approach (plasma HHG) may offer as a method for both fundamental science and practical applications.

The main concern for coherent XUV pulses is the typically rather weak strength and low generation efficiency. Many applications require substantially stronger XUV brightness. To achieve higher intensity of generating harmonics, laser-plasma interaction could have been exploited as a promising alternative. The reasons to think so are presented above [see (a)–(g)]. Moreover, low-density and low-ionized plasma can withstand high laser fields driving the harmonics.

Though not applicable presently for micro-lithography as one of most promising fields of applications of short wavelength sources, coherent XUV sources based on gas and plasma HHG could be useful elsewhere in semiconductor manufacturing because they enable high-resolution imaging. The ability to focus this radiation into a small spot (of the order of 10 to 50 nm) makes it possible to ablate a sample, such as a microorganism, and study its chemical composition at different spots for the goals of biology and medicine. Other coherent XUV research applications include patterning, imaging, photochemistry and microscopy.

Plasma HHG has opened new doors in many unexpected areas of light-matter interaction. Apart from being an alternative method for generation of coherent XUV radiation, it has proved to be a powerful tool for various spectroscopic and analytical applications. The application of doubly charged ions for high-order harmonic generation showed a promising extension of the cut-off photon energy of plasma harmonics, without having to rely on few-cycle driving pulses. As was shown in the case of low- and high-order harmonics, the conversion efficiency can be strongly enhanced in the vicinity of the resonances of atomic or ionic systems. This enhancement of a single harmonic was first demonstrated during experiments using indium plasma and then confirmed in other plasmas. For laser-generated plasmas, a large variety of materials can be employed, thereby increasing the chance to match such resonances with the radiation of Ti:sapphire lasers. Furthermore, it was shown in gas HHG studies that the two-color pump profitably enhances the high-order harmonic intensity and significantly influences the output and properties of the harmonic spectrum generation in noble gases. For plasma harmonics, where a two-color pump technique has recently been introduced, this is a new approach for the nonlinear spectroscopy of the numerous ionic transitions of ablating materials possessing high oscillator strengths.

Plasma harmonic studies have shown that the enhanced high-order harmonics can also be generated from ablated nanoparticles, which opens the prospects for applications of the local field enhancement, use of broad plasmonic resonances in the XUV and efficient recombination processes for plasma HHG. As a highly interesting perspective an increase in the harmonic output by quasi-phase-matching in specially prepared plasmas has been developed and offered larger variety of applications compared with gas harmonics concept.

Among the achievements that emerged in recent years one can admit the comparative theoretical and experimental studies of silver plasma harmonic cut-off, new findings in fullerenes' high-order non-linearities, single sub-femtosecond harmonic generation in manganese plasma using few-cycle pulses, comparative research of plasma and gas media for efficient HHG, temporal characterization of plasma harmonics, generation of continuum plasma harmonics, stabilization of harmonic yield over one million laser shots on the rotating targets, generation of high-order harmonics using picosecond driving pulses, various applications of 1 kHz lasers for plasma HHG to increase the average power of converted radiation, analysis of coherence properties of plasma harmonics, use of double-pulse technique for plasma HHG, demonstration of the quantum path interference of the long and short trajectories of electrons in plasma HHG experiments, etc. All those findings substantially pushed ahead our knowledge on the peculiarities of plasma media through the analysis of their high-order nonlinear optical characteristics.

Similarly to the developments of MIR pump based gas HHG, the same approach reveals numerous interesting peculiarities during plasma HHG. One can expect that various approaches for further amendments of plasma HHG will be examined. Among them the following approaches should be mentioned: harmonic generation in plasma using the two-color pump in the case of commensurate and incommensurate wavelength sources in the mid-IR and ultraviolet ranges, analysis of the influence of molecular orientation on the MIR-induced harmonic output from molecular plumes, development of ablation-induced HHG spectroscopy of various organic materials using MIR pulses, analysis of orientation-induced nonlinear optical response of large ablated molecules and clusters, and the time-resolved pump-probe analysis of the complex plasmas containing various molecules. Motivated largely by the vast potential of attosecond science, the development of ultraintense few-cycle and CEP-stable sources based on gas HHG concept has intensified, and it was recognized that coherent soft x-ray radiation could be generated when driving HHG with long wavelength sources [10–12].

There are a few main aspects, which distinguish the proposed MIR-induced approach with regard to the already reported studies of plasma harmonics and show the competetiveness of the plasma harmonic concept over the gas harmonic one (i.e. resonance enhancement, application of clustered media, quasi-phase-matching of specifically modulated plasma plumes, etc.) for generation of ultrashort pulses in the XUV region. These aspects are discussed throughout this book. The key enabling technology for this research is the capability of laser ablation to put atoms and molecules from solids in gas phase at densities sufficient for HHG attosecond spectroscopy measurements. In particular, characterization of the MIR-induced XUV ultrashort pulses generated in the plasma plumes during optimized HHG and application of attosecond pulses for nonlinear spectroscopy of ablated materials could be considered as the following steps of MIR-induced plasma HHG studies.

It has already been underlined that it would be interesting to analyse the application of MIR (1000–5000 nm) radiation to study the dynamics of the nonlinear optical response of ablated molecular structures compared with commonly used Ti: sapphire lasers for plasma HHG, including the studies of extended harmonics at a comparable conversion efficiency with shorter wavelength laser sources, and a search for new opportunities in improvement of the HHG conversion efficiency in the mid-IR range, such as the application of clustered molecules. Thus the motivation for writing this book is to show the most recent findings of various new schemes of the application of MIR pulses for HHG in laser-produced low-ionised, low-density plasma plumes, which could be dubbed for simplicity as 'plasma harmonics'. The use of any elements of the periodic table, as well as thousands of complex samples that exist as solids greatly extends the range of materials employed, whereas only a few rare gases are typically available for gas HHG. The exploration of practically any solid-state material through the nonlinear spectroscopy comprising laser ablation and harmonic generation can be considered as a new tool for materials science. Thus the MIR pump based laser-ablation-induced high-order harmonic generation spectroscopy can be considered a new method for the study of materials and one of most important applications of plasma HHG.

This book is organized as follows. To emphasize previous studies of MIR induced harmonics we discuss in Chapter 1 various applications of mid-infrared sources for the study of the high-order nonlinear optical processes in gases. Here we analyze the architecture of mid-infrared lasers used for harmonic generation in gases and the HHG harmonic generation in gases using MIR pulses. In particular, infrared two-color multicycle laser field synthesis for generating an intense attosecond pulse, attosecond nonlinear optics using gigawatt-scale isolated attosecond pulses, generation of coherent radiation in the water window region at 1 kHz repetition rate using a mid-infrared pump source, and various approaches in the HHG using MIR pulses are discussed. In Chapter 2, the principles of HHG in laser-produced plasmas using MIR pulses are analyzed. Among them, we present the overview of the HHG graphite plasma plumes using 800 nm and MIR laser pulses, the MIR- and 800-nm-induced HHG in the laser plumes contained the components of DNA, and the HHG in fullerenes using few- and multi-cycle pulses using different MIR and visible wavelengths. Chapter 3 is dedicated to the resonance-induced enhancement of harmonics using tunable MIR radiation. We discuss the resonance enhancement of harmonics in various metal plasmas using tunable mid-infrared pulses, including the experimental studies and theoretical analysis of resonance-enhanced harmonic spectra from Sn, Sb, and Cr plasmas, show the enhancement of tunable harmonics in the indium plasma using single- and two-color mid-infrared fields and present the theory of resonance enhancement in this medium, and analyze the resonance enhancement of harmonics in laser-produced Zn II and Zn III containing plasmas including modification of harmonic spectra at excitation of neutrals and doubly charged ions of Zn. The quasi-phase-matching in plasmas using mid-infrared pulses is discussed in Chapter 4. In particular, we show the application of MIR pulses for quasi-phase-matching of high-order harmonics in silver plasma, present the overview of early studies of quasi-phase-matching, and describe the theory of QPM. Here we also present the studies of the on- and off-axis quasi-phase-matching of the harmonics generated in the multi-jet laser-produced plasmas, including the description of the problem and the analysis of the on- and off-axis conditions of QPM, and compare the influence of the micro- and macro-processes on the HHG in laser-produced plasma. Chapter 5 contains various applications of MIR HHG approach in different plasma plumes. Among them the ablation of boron carbide for the HHG of ultrafast pulses in laser-produced plasma, two-color HHG in plasmas, including the microscopic and macroscopic responses, which influence the harmonic generation efficiency depending on the properties of the plasma particles, and the high-order sum and difference frequencies generation using tunable two- and three-color commensurate and incommensurate MIR pumps of graphite plasma are presented and discussed. Finally, the discussion about the perspectives of this approach, mainly aimed in attosecond pulses generation and spectroscopy, summarizes this book.

Chapter 1

Applications of Mid-infrared Sources for the Study of the High-order Nonlinear Optical Processes in Gases

1.1. Architecture of Mid-infrared Lasers Used for Harmonic Generation in Gases

1.1.1. *Various schemes of mid-infrared sources*

The development of coherent light sources with emission of short-wavelength ultrashort pulses using mid-infrared (MIR) sources is currently undergoing a remarkable revolution [1]. The MIR spectral range has always been of tremendous interest, mainly to spectroscopists, due to the ability of MIR light to access rotational and vibrational resonances of molecules, which give rise to superb sensitivity upon optical probing [2–4]. Previously, high energy resolution was achieved with narrow-band lasers or parametric sources, but the advent of frequency comb sources has revolutionized spectroscopy by providing high energy resolution within the frequency comb structure of the spectrum and at the same time broadband coverage and short pulse duration [5–7]. Such carrier-to-envelope-phase (CEP) controlled light waveforms, when achieved at ultrahigh intensity, give rise to extreme effects such as the generation of isolated attosecond pulses in the vacuum to extreme ultraviolet range (XUV) [8].

Motivated largely by the vast potential of attosecond science, the development of ultraintense few-cycle and CEP-stable sources has intensified [9], and it was recognized that coherent soft x-ray radiation could be generated when driving high-order harmonic generation (HHG) with long wavelength sources [10–12]. Recently, based on this concept, the highest waveform controlled soft x-ray flux [13] and isolated attosecond pulse emission at 300 eV [14] was demonstrated via HHG from a 1850 nm, sub-two-cycle source [15]. Within strong-field physics, long wavelength scaling may lead to further interesting physics such as the direct reshaping of the carrier field [16], scaling of quantum path dynamics [17], the breakdown of the dipole approximation [18], or direct laser acceleration [19]. The experimental development of long wavelength light sources therefore holds great promise in many fields of science and will lead to numerous applications beyond strong field physics and attosecond science.

There are many configurations of MIR lasers depending on the characteristics required for different needs. Particularly, the first MIR optical parametric chirped pulse amplifier (OPCPA) operating at a center wavelength of $7 \mu m$ with output parameters suitable already for strong-field experiments has been described in [20]. It is the first demonstration of OPCPA using a $2 \mu m$ laser pump source, which enables the use of nonoxide nonlinear crystals with typically limited transparency at $1 \mu m$ wavelength. This new OPCPA system is all-optically synchronized and generates 0.55 mJ energy, CEP-stable optical pulses. The pulses supported a sub-four-cycle pulse duration and are currently compressed to fewer than eight optical cycles due to uncompensated higher-order phase from the grating compressor, which will be addressed in the future.

Recent impressive progress in generation of high-power ultrashort pulses in the MIR [21] opens new horizons in ultrafast optical science and technologies, allowing the generation of unprecedentedly broad harmonic spectra [22], enabling lasing in laser-induced filaments [23], and revealing unusual phenomena and unexpected properties of materials in the MIR range [24, 25].

One of the key challenges of ultrafast optical science in the MIR range that still needs to be addressed is finding the routes toward efficient generation of few- and single cycle MIR pulses [26]. Several promising techniques have been proposed to confront this challenge. Generation of subcycle pulses at a central wavelength of $3.9 \mu m$ has been demonstrated using filamentation of a two-color laser field consisting of a powerful ultrashort near-IR pulse and its second harmonic [27]. The throughput of this method of pulse compression, however, is inevitably very low. Optical parametric amplification using periodically poled lithium tantalate has been shown to enable few-cycle pulse generation in the wavelength region from 2 to 5 μm [28]. Self-compression of MIR pulses with a central wavelength of 3.1 μm to a sub-three-cycle pulse width has been demonstrated using an anomalously dispersive solid dielectric [29, 30].

However, pulse compression in the regime of anomalous dispersion is usually not easily scalable to higher energies. Theoretical studies show that high-power sub-100-fs MIR pulses available from optical parametric chirped-pulse amplification can be compressed to few-cycle pulse widths using filamentation and gas-filled hollow waveguides [31]. These methods, however, are limited to the MIR range centered at approximately 4 μm, where an efficient solid-state source of high-power sub-100-fs MIR pulses is available.

The authors of [26] have identified a physical scenario whereby freely propagating MIR pulses can be compressed to pulse widths close to the field cycle. This physical scenario involves self-focusing-assisted spectral broadening in a normally dispersive, highly nonlinear semiconductor material, followed by pulse compression in the regime of anomalous dispersion, where the dispersion-induced phase shift is finely tuned by adjusting the overall thickness of anomalously dispersive components. This approach is shown to enable the generation of tunable few-cycle pulses in the wavelength range from 4.2 to 6.8 μm, yielding field waveforms with a pulse width of less than 1.5T_0, where T_0 is the field cycle, and a peak power up to 60 MW at a central wavelength of 5.9 μm.

However, the most used MIR sources for HHG comprise those based on commercially available technologies. In this section, we analyze the generation of 10 mJ, 5-cycle pulses at 1800 nm (30 fs) at 100 Hz repetition rate using an optical parametric amplifier pumped by a high energy titanium-sapphire (Ti:sapphire) laser system (total energy of 23 mJ for signal and idler). This is the highest reported peak power (0.33 TW) in the infrared spectral range. This high-energy long wavelength laser source is well suited for driving various nonlinear optical phenomena such as HHG for high flux ultrafast soft X-ray pulses.

1.1.2. *Commercially available source of mid-infrared radiation*

Since the discovery of the laser in 1960, a large variety of nonlinear optical phenomena have been discovered including optical parametric amplifier (OPA) [32, 33]. Nowadays, OPAs are widely used in combination with Ti:sapphire laser systems for generating tunable fs pulses in the infrared (IR) range with the signal and idler covering the spectral range of ~1.2 to ~2.5 μm [34], to even longer wavelengths in the MIR via difference frequency generation (DFG) between signal and idler [35], but also to shorter ones through the use of OPA in combination with harmonic generation and sum frequency generation nonlinear optical processes [36]. This provides access to fully synchronized femtosecond (fs) pulses from ~ 200 nm to 20 μm, offering a unique tool for a wide variety of time-resolved applications.

In recent years, high peak power IR fs lasers have become highly attractive sources as several nonlinear optical processes, in particular, the ones in the strong field regime, scale advantageously with increasing laser wavelength as the ponderomotive energy of free electrons is $I \propto \lambda^2$. Intense IR fs pulses are ideal for: (i) scaling the process of HHG in the soft X-ray spectral range with the breaking record of 1.6 keV photon energy [22], (ii) high spatial resolution molecular imaging using laser induced electron diffraction [37], (iii) generation of intense single-cycle THz pulses through driving plasma emission in a two-color laser field [38], (iv) generation of sub-100 fs hard X-ray pulses using laser driving plasma emission from a copper target [39], and (v) generation of multi-octave light sources [29].

Those advances have been made possible as intense research activities are carried out to develop laser sources delivering high peak power IR pulses, from multi-cycles to near single-cycle pulse duration. This major effort can be divided into two categories: OPA and OPCPA [40]. Using OPA, weak few-cycle laser pulses have been generated using various approaches [41], whether through the OPA of broadband seed pulses obtained via intrapulse DFG [42], or by spectral broadening through nonlinear propagation in hollow core fiber (HCF) followed by dispersion compensation either using chirped mirrors [43] or linear propagation in the anomalous dispersion regime [44]. For multi-cycle duration, Takahashi *et al.* used OPA to generate 1.4 μm 40 fs pulses carrying 7 mJ at a repetition rate of 10 Hz, which were at the time the highest peak power (0.175 TW) pulses in the IR spectral range [45] using conventional OPA scheme.

With OPCPA, several research teams have demonstrated the millijoule of energy per pulse with few-cycle duration in the IR spectral range [46, 47]. In addition to those developments, high peak power OPCPA at 3.9 μm (0.1 TW) [21] has been the key technology for reaching 1.6 keV using HHG [22] but is still limited to 20 Hz repetition rate. As a new perspective, frequency domain OPA (FOPA) has recently been developed to amplify few-cycle IR pulses, by tailoring the phase matching conditions over near an octave of spectral bandwidth by using multiple OPA crystals installed in the Fourier plane of a 4f setup [48], with 0.1 TW peak power. Finally, both OPCPA and FOPA techniques will strongly benefit from the rapid development of high average power ytterbium picosecond laser systems [47, 49].

Despite those great advances, conventional OPA remains the simplest approach to generate high energy IR laser pulses. Here we discuss a straightforward scheme reported in [50] based on OPA to generate 10 mJ, 5-cycle pulses at 1.8 μm (30 fs), which is the highest peak power (0.33 TW) reported in the IR spectral range. This simple approach for generating high energy IR laser pulses presents a major technical advantage with respect to OPCPA as there is no need to stretch and compress the seed pulses. It is also free of pulse shapers.

A scheme of the experimental setup used is shown in Fig. 1.1. The IR laser source is based on a conventional Ti:sapphire laser system providing up to 80 mJ per pulse (80 mJ/p) at a repetition rate of 100 Hz. A small fraction of this beam (5 mJ per pulse) is compressed

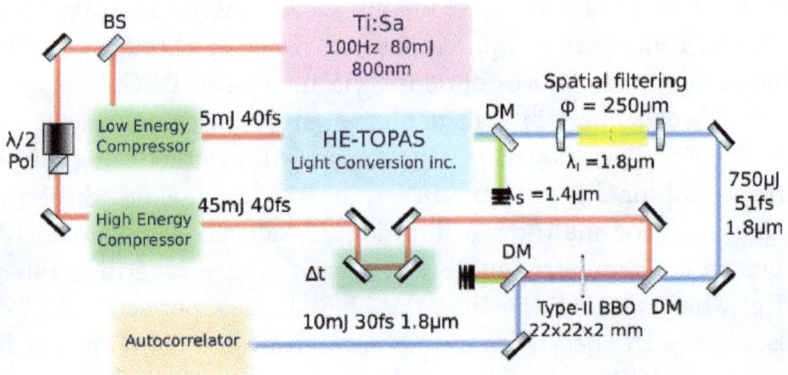

Fig. 1.1. Experimental setup for the generation of high peak power IR pulses; BS: Beam splitter; DM: Dichroic mirror; Pol: Polarizer; φ: Pinhole diameter used for spatial filtering; Autocorrelator: Home build all reflective second harmonic generation autocorrelator; λ_I: central wavelength of the idler; λ_s: central wavelength of the signal. Reproduced from [50]. Copyright 2015. AIP Publishing LLC.

to 45 fs by the first grating compressor (low energy compressor, see Fig. 1.1) and used to pump a white light seeded high-energy OPA (HE-TOPAS, Light Conversion, Inc.). Briefly, a broadband seeding signal is generated through white light generation in a sapphire window.

The signal is then parametrically amplified around 1.45 μm by the 800 nm pump in a Beta Barium Borate (BBO) crystal. The idler generated in the first amplification stage (1.8 μm) is then amplified by two additional OPA stages, which are also based on BBO crystals pumped by 800 nm. All BBO crystals are cut at 27° for type-II phase matching. At the output of the HE-TOPAS, a 1.1 mJ/p of signal (1.45 μm) and a 0.9 mJ/p of idler (1.8 μm) were obtained. This corresponds to a conversion efficiency of 40%. A dichroic mirror is then used to separate the signal and idler beam. The idler is spatially filtered by focusing it through a 250 μm diameter pinhole. The pinhole is installed inside a primary vacuum chamber to avoid any nonlinear effects in air. The transmission of the spatial filter is typically around 75%–80%. The CEP of the idler is passively stable. An intensity fluctuation of less than 1%–2% root mean square, and 6% peak-to-peak, is typically achieved at the output of the spatial filter.

The remaining energy of the Ti:sapphire system is also compressed to 45 fs using another grating compressor (high energy compressor, see Fig. 1.1). The amount of energy going into the high-energy compressor is controlled with a motorized half-wave plate located in front of a polarizer. This high energy pump beam is then used to amplify the idler ($1.8\,\mu$m) coming from the HE-TOPAS in a large type-II phase matching BBO crystal ($22\,$mm $\times$ $22\,$mm $\times$ $2\,$mm, cut at $30°$) which correspond to the last (external) parametric amplification stage. The amplified $1.8\,\mu$m beam is no longer CEP stable at the output of this OPA stage because of the temporal jitter between the high- and the low-energy compressor [51]. This jitter is rather small as the energy stability measured at the output of the high energy amplification stage is similar to the one at the output of the HE-TOPAS. Nonetheless, a small temporal jitter reshapes the amplified envelope, thus moving its peak position compared to the electromagnetic field resulting in the loss of CEP stability. The easiest solution to avoid this problem will be to use only one compressor at the cost of inconvenience as this will limit the capability to vary the pump energy.

A crucial step towards the generation of $10\,$mJ/p class IR beamline is the parametric amplification of the commercially available HE-TOPAS idler output. The crystal used for this amplification stage is a BBO crystal with cutting angle $30°$ to satisfy type-II phase matching condition between pump, signal, and idler. The IR spectra at the output of the HE-TOPAS ($750\,\mu$J/p after the spatial filter) and after the home built high energy OPA stage ($10\,$mJ/p), are shown in Fig. 1.2(a). The idler ($1.8\,\mu$m) spectra at the output of the OPA has 95 nm of bandwidth corresponding to Fourier transform limited pulses (FTL) of 50 fs. Comparing the spectra before and after the last amplification stage, one can see that the gain is not homogeneous over the entire spectral range. To achieve this inhomogeneous spectral gain, the BBO crystal angle was slightly tuned to favor the phase matching condition being fulfilled for the red part of the spectra. The

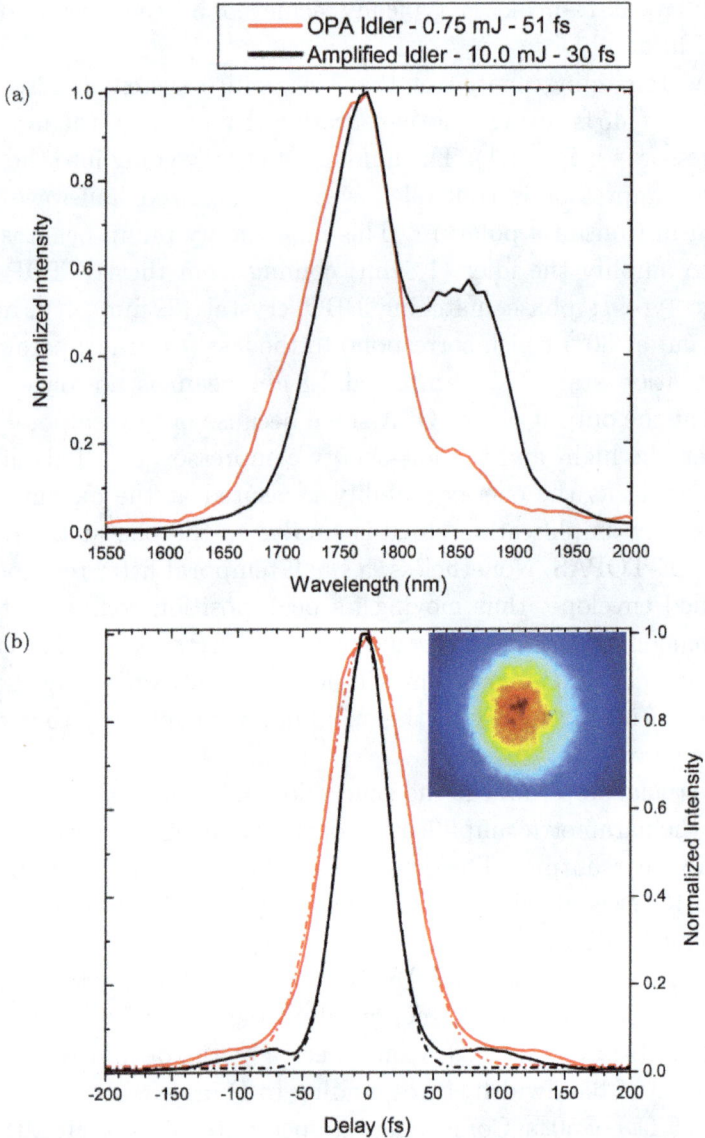

Fig. 1.2. Pulse characterization. In (a): the spectrum of the unamplified OPA Idler in red and the spectrum of the amplified Idler ($1.8\,\mu$m) in black. In (b): the autocorrelation trace of the OPA output in red and of the amplified IR pulses ($1.8\,\mu$m) in black. The dotted-dashed curves in red and in black are Gaussian fit of the autocorrelation trace of the OPA Idler and the amplified idler, respectively. Reproduced from [50]. Copyright 2015. AIP Publishing LLC.

amplified idler spectra supported FTL pulses of 28 fs. The measured 1.8 μm pulse duration was 51 fs before amplification and 30 fs after, values close to FTL pulse duration.

The spatial beam-intensity profile is characterized using a Si-based CCD camera. As the photon energy of the 800 nm pump (E = 1.55 eV) is larger than the Si detector band gap (E = 1.11 eV), the pump beam diameter is straightforwardly imaged on the chip through single photon excitation. The IR beam centered around 1.8 μm, corresponding to photon energy of 0.67 eV, is detected via a two photon excitation of the Si detector. Because of this nonlinear based detection, the recorded spatial beam-intensity profile needs to be deconvoluted assuming a second-order process. The unamplified OPA idler beam diameter is 16 mm. The pump beam diameter (800 nm) was adjusted to 20 mm, to be slightly below the intensity threshold where self-phase modulation starts to appear in the high energy OPA stage while pumping with 45 mJ/p of Ti:sapphire laser. The amplified 1.8 μm beam diameter is 19 mm. All the aforementioned diameter values are at $1/e^2$ of intensity. The high spatial quality of our amplified idler beam is presented in Fig. 1.2(b).

In Fig. 1.3, the results of the characterization of the gain in the high energy OPA stage as a function of the seed (1.8 μm) and the pump (800 nm) energies are presented. The pump energy was varied from 8 mJ/p to 45 mJ/p for seed levels of 750 μJ/p and 250 μJ/p, respectively. One can see the linear progression of the output energy of the amplified idler beam.

With 750 μJ/p, the amplified 1.8 μm beam energy increases from 1.5 mJ/p (8 mJ/p for the pump) to 10.0 mJ/p (45 mJ/p for the pump), whereas for the 250 μJ/p seed the maximum output energy is 8 mJ/p. They have also characterized the amplified 1.8 μm output energy while keeping the pump level at 45 mJ/p and varying the seed level for 150 μJ/p, 250 μJ/p, 350 μJ/p, and 750 μJ/p (see the inset of Fig. 1.3) with the same duration. When seeding with an intensity of 750 μJ/p and pumping with 45 mJ/p, they have obtained 10.0 mJ/p of idler (1.8 μm) and 13.0 mJ/p of signal (1.45 μm) for a total of 23 mJ. This corresponds to a remarkable conversion efficiency of 50%.

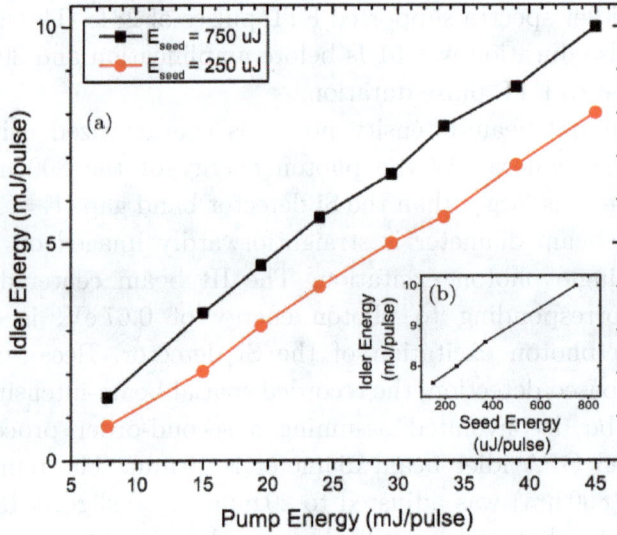

Fig. 1.3. (a) Amplified idler energy per pulse as a function of pump energy (seed level fixed). In black, the seed energy is fixed at $750\,\mu J/p$ and in red at $250\,\mu J/p$. (b) Amplified idler energy per pulse as a function of seed energy (pump level fixed at $45\,mJ/p$). Reproduced from [50]. Copyright 2015. AIP Publishing LLC.

1.2. High-order Harmonic Generation in Gases Using MIR Pulses

Attosecond science has emerged as an important research area of ultrafast phenomena within the past decade. Owing to its numerous successes, attosecond science has created important new knowledge of the fundamental science of the interaction between electrons and photons. Recently, some key research topics have been identified, which are expected to make major breakthroughs for the next attosecond frontiers, such as the MHz high-repetition high-order harmonic frequency combs for the joint frontier of precision spectroscopy and ultrafast science, attosecond-pump/attosecond-probe experiments for observing and controlling electronic processes in atomic and molecular physics, and nonlinear science occurring on the attosecond time scale.

To seriously tackle the above interesting research topics, one of the most important issues is the development of high-power isolated attosecond pulses (IAPs) and/or attosecond pulse trains. IAPs and attosecond pulse trains are produced using HHG in gases. To date, high-power attosecond pulse trains have been successfully generated owing to research on HH energy scaling using a loose-focusing geometry. Attosecond pulse trains with sufficiently high energy in the microjoule range with a high conversion efficiency on the 10^{-4} level were generated, which are intense enough for implementing some applications, such as nonlinear attosecond optics without the assistance of laser pulses, single-shot femtosecond holography and an external injector for a seeded free electron laser (FEL) in the XUV region.

In this section, we present a few concepts of harmonics generation in the gaseous media allowing the generation of ultrashort (attosecond) pulses.

1.2.1. *Infrared two-color multicycle laser field synthesis for generating an intense attosecond pulse*

To achieve a breakthrough in attosecond science [1] for nonlinear optics and other applications, one of the most important issues is the development of high-power attosecond pulse sources. The progress of high-order harmonic generation techniques has resulted in the creation of IAPs [8, 52–55] and attosecond pulse trains (APTs) [56, 57]. High-power APTs have been successfully generated as a result of research on harmonic energy scaling using a loosely focusing geometry [58]. Meanwhile, various methods have been proposed for creating an IAP, such as the use of a few-cycle IR pulse as a driving laser [8, 52, 53], polarization gating [55], double optical gating (DOG) [59], and so forth [60–63]. Although the shortest pulse duration of IAPs attained is 80 as [64], the output energy is still not sufficient to induce nonlinear phenomena because the pump pulse energy is typically limited to a few mJ owing to the requirements of sophisticated laser technology such as a few-cycle pulse duration and CEP stabilization. The scalability of pump laser energy is of paramount importance for the development of intense IAP sources. It is, however, still difficult to apply high-power conventional femtosecond laser technology to increase the power of IAPs.

Here we discuss the studies of the generation of a continuum high-order harmonic spectrum by mixing multicycle two-color (TC) laser fields with the aim of easily obtaining an intense IAP [65]. The authors of this study utilized a TC IR driver source to greatly reduce the requirements for the pump laser system used for generating an IAP. A similar theoretical proposal has been reported by other groups [66,67]; essentially, they assumed that a CEP-stabilized pulse is used for the driving field. As mentioned above, the requirement of CEP stabilization is one of the factors hindering the generation of an intense IAP by a TW-class high-power laser system. To meet this requirement for laser technology, they optimized and demonstrated the creation of an IAP with non-CEP-stabilized multicycle laser pulses. By adjusting the wavelength of the supplementary IR pulse, they suppressed the multiple pulse burst and successfully generated an IAP.

To demonstrate and verify the proposed concept, they have carried out a two-color-induced high-order harmonic generation (TC-HHG) experiment using a high-energy IR source [45, 68] based on OPA. Signal pulses obtained from the OPA scheme are used for the supplementary IR pulse. Since the external IR seed pulses for the final OPA stage are produced by white-light continuum generation, the CEP of the IR pulse is determined by the 800 nm pulse and preserved in the amplified signal pulse [69]. The durations of the 800 nm pump pulse and signal IR pulse are 30 and 40 fs, respectively. The TC pulses are focused using an R = 400 mm concave mirror. The target gas is supplied by a 2-mm diameter synchronized supersonic gas jet operating at 10 Hz. The generated harmonics illuminate the slit of a spectrometer.

To investigate the generation efficiency of this TC scheme, they measured the harmonic spectrum by accumulating laser shots. Figure 1.4(a) shows the average Ar harmonic spectrum for 800 laser shots driven by a TC laser field (800 nm + 1300 nm) at a focused intensity of 1.15×10^{14} W cm^{-2}. The harmonic yield is optimized

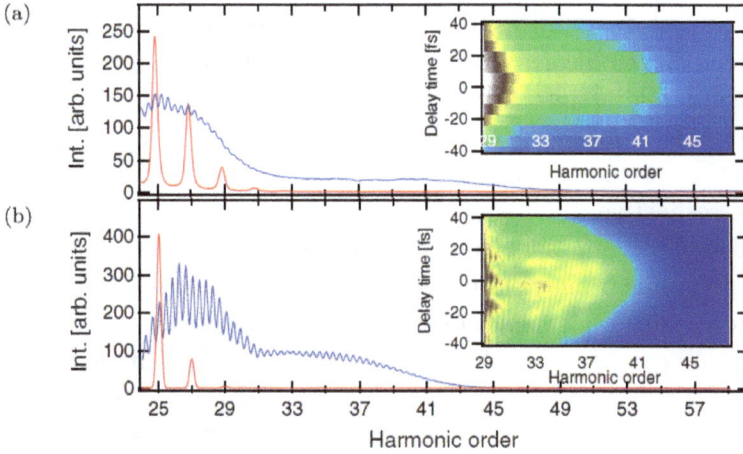

Fig. 1.4. (a) Experimentally obtained average Ar harmonic spectrum for 800 laser shots. Red line: SC field; blue line: TC field. (b) Calculated harmonic spectra. Red line: intensity of 1.0×10^{14} W cm^{-2} for 800 nm pulse with 30 fs duration; blue line: average harmonic spectrum from $-\pi/2$ to $+\pi/2$ rad at an intensity of 1.15×10^{14} W cm^{-2} for TC field. Insets show delay time dependence of HH spectra near cutoff. Reproduced from [65] with permission from American Physical Society.

by adjusting the focusing point to the gas jet and by varying the backing pressure of the gas jet. The harmonic yield gradually increases as the gas pressure increases, and exhibits a quadratic dependence on the gas backing pressure. The inset of Fig. 1.4(a) shows the delay time dependence of the harmonic spectrum near the cutoff region. When the delay time between the 800 and 1300 nm pulses is sufficiently large, the harmonic spectrum has a discrete structure, which is the same as that obtained from a single-color (SC) field with a wavelength of 800 nm (red profile). The cutoff order gradually increases as the delay time is decreased to 0 fs (blue profile), which is caused by the increase in intensity due to the mixed laser field. When reducing the intensity of the supplementary IR pulse while maintaining the intensity of the 800 nm pulse, the continuum harmonic spectrum gradually disappears. Note that the harmonic intensity of the TC laser pulse is comparable to that of the SC laser field. Taking previous experimental results into account [58], one can expect to generate microjoule XUV energy with conversion efficiency of 10^{-5} by using 5-m focusing condition. To verify the obtained harmonic features, they calculated the average harmonic spectrum taking into account the effects of propagation and the atomic structure of Ar [70]. The blue and red lines in Fig. 1.4(b) show the calculated harmonic spectrum generated by TC and SC fields, respectively.

The harmonic spectrum of the TC field corresponds to the average profile from $-\pi/2$ to $+\pi/2$ rad. The inset of Fig. 1.4(b) shows the delay time dependence of the harmonic spectrum near the cutoff region. As can be seen in Fig. 1.4(b), the cutoff is gradually extended as the delay time is decreased to 0 fs, and the harmonic spectrum has a dense structure. The calculated spectra are in reasonable agreement with the experimental results.

To analyze the fine structure of the spectrum, they acquired a single-shot spectrum for all laser shots by synchronizing the measurement system. Figure 1.5(a) shows a 2D single-shot harmonic spectrum imaged on a microchannel plate generated by the TC field (800 nm + 1300 nm). A dense harmonic spectrum appears in the

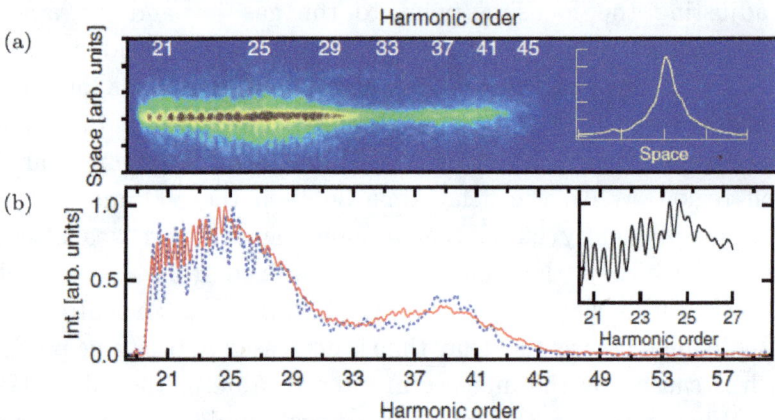

Fig. 1.5. (a) 2D single-shot harmonic image driven by the TC field. The laser parameters are the same as those in Fig. 1.4(a). (b) Single-shot 1D harmonic spectrum driven by the TC field. The insets in the top and bottom panels correspond to the spatial profile of cutoff harmonics (38th to 45th) and an enlargement of the spectrum from the 21st to 27th harmonic order. Reproduced from [65] with permission from American Physical Society.

lower-order region [see the inset of Fig. 1.5(b)]. This dense harmonic spectrum can be considered as high-order sum and difference frequency generation (SFG, DFG) [71], which is also reproduced by the numerical calculation [see Fig. 1.3(b)]. Each harmonic component is assigned to SFG and/or DFG between 800 and 1300 nm. Since the resolution of the spectrometer is not sufficiently high to perfectly resolve each order in the dense spectrum, the modulation depth of the spectrum has a shallow structure. However, the dense components disappear as the harmonic order reaches the cutoff region, and a continuum harmonic spectrum can clearly be seen (33rd to 45th order). The inset of Fig. 1.5(a) shows the measured far-field spatial profile of the harmonic beam (38th to 45th order). In this TC scheme, they obtained a high-quality beam with a Gaussian-like profile because the laser intensity is fixed to less than the ionization threshold of the harmonic medium. When they changed the wavelength of the IR pulse from 1300 to 1400 nm while maintaining the focused intensity, the measured spectrum maintains its continuum structure [blue line in Fig. 1.5(b)] as predicted on the basis of their concept.

As already discussed, the harmonic intensity in the cutoff region is affected by the CEP value. To evaluate the CEP effect, they simultaneously measured the harmonic spectrum and corresponding CEP for each laser shot using a nonlinear interferometer, which records the interference fringes between the pump pulse spectrally broadened by white-light generation in a sapphire plate and the second harmonic of the idler pulse near 1000 nm. The intensity ratio of the mixed IR field is slightly increased from 0.15 to 0.25 when the main field is fixed at 1×10^{14} W cm^{-2}. As shown in the inset of Fig. 1.6(b), the measured CEP value randomly changes for each laser shot. Figure 1.6(a) exhibits the typical cutoff spectra at CEP of -0.7 and 0.5 rad. For CEP values from -1.0 to 0.0 rad, they observed an intense harmonic spectrum in the cutoff region. To evaluate the measured results, the harmonic spectra were calculated under the experimental conditions. The dashed red and blue lines in Fig. 1.6(a) correspond to the calculation results, which are in good agreement with the experimental results. Figure 1.6(b) shows the measured variation of cutoff intensity (44th to 49th) as a function

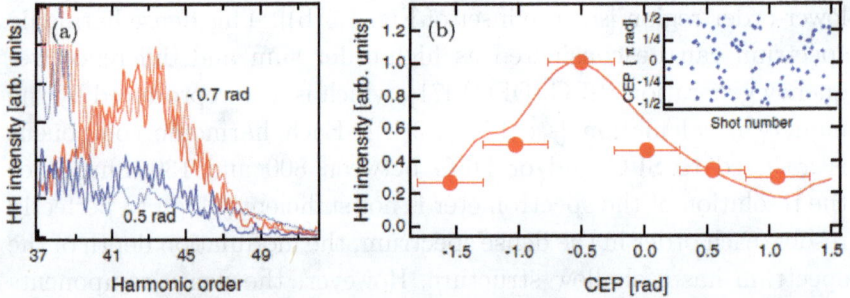

Fig. 1.6. Measured (solid line) and calculated (dashed line) harmonic spectra at CEP of −0.7 and 0.5 rad. (b) Variation of measured cutoff HH intensity as a function of CEP. The red curve shows the calculated intensity variation. Inset: History of the CEP value. Reproduced from [65] with permission from American Physical Society.

of CEP, which is also in reasonable agreement with the calculation result (red line). This CEP dependence clearly shows the possibility of generating IAPs from multicycle laser pulses by synthesizing TC laser fields.

1.2.2. *Attosecond nonlinear optics using gigawatt-scale isolated attosecond pulses*

The output energy of IAPs is still not sufficient for various applications, although the shortest pulse duration achieved is sub-100 as. The widespread application of IAPs has been limited because of the low photon flux and the complexity of the laser systems required to produce IAPs [72]. At present, the highest IAP energy is less than a few tens of nanojoules at a 27-eV photon energy. The highest conversion efficiency demonstrated is only 10^{-6} to 10^{-5}, even if Xe gas is used. Consequently, the production, characterization and application of high-energy IAPs are still under active investigation.

Here we analyze a harmonic generation method suitable for nonlinear pump/probe experiments that demonstrates the capability of time-resolving attosecond dynamical processes with IAPs, which was reported in [73]. The generation scheme proposed in this paper is robust and straightforward for scaling up the harmonic energy, which is based on an infrared two-color laser field synthesis [65] and the energy-scaling method. By carefully designing the HHG configuration, they obtained IAPs with a record energy of up to 1.3 mJ at 30 eV, thus demonstrating a 4100-fold energy enhancement compared with previous reports [74]. The conversion efficiency achieved is improved to 1.1×10^{-4}, owing to favorable phase-matching technique. Even though their TC method leads to weak multiple burst emissions [73] occurring for specific combinations of the CEP and relative phase, they nevertheless observed the nonlinear signal corresponding to IAPs, because the nonlinear process automatically selects predominantly the contributions from the most intense IAPs generated for certain favorable phase combinations. They presented the direct temporal characterization of an IAP by a second-order autocorrelation (AC) measurement using nonlinear phenomena in N_2 molecules. The AC measurement is one of the simplest applications of a nonlinear pump/probe experiment. The measured AC traces indicate not only the temporal duration of the interacting IAPs but also provide evidence for the nonlinear interaction with the target medium. From the 500-as pulse duration determined by the AC measurement, the IAP's peak power can be evaluated to be 2.6 GW with weak satellite pulses at the generation point. Furthermore, in those TC-HHG experiment, they created a mid-plateau region exhibiting a quasi-continuous spectrum with an energy of 10 mJ. From the AC measurement of this quasi-continuous spectrum at the mid-plateau, they evaluated that a quasi-IAP with pulse duration of 375 as with $\sim$0.27 intensity satellite pulses at $\pm$6.7 fs dominated the nonlinear interaction with the N_2 target.

800-nm and infrared pulses are generated starting from a home-built Ti:sapphire laser system (30 fs, 150 mJ, 800 nm, 10 Hz; Fig. 1.7). Infrared pulses with a 40-fs duration are produced by a white-light-seeded high-energy optical parametric amplifier [45]. To correct for the different focal lengths of the 800-nm and infrared pulses, both pulses were focused using two separate focusing lenses (f = 4.5 m for 800 nm, f = 3.5 m for infrared). The two beams are collinearly combined with a dichroic beam splitter and delivered into the target chamber through a thin CaF_2 window. The focused beams pass through the interaction gas cell, which has two pinholes on

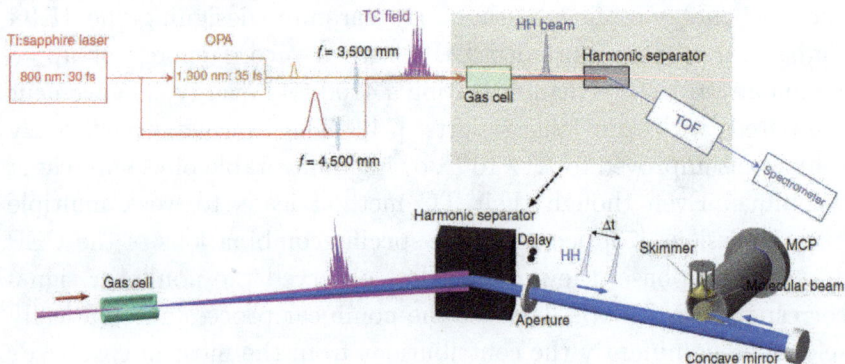

Fig. 1.7. Schematic figure of the HHG setup and the second-order autocorrelator. The AC setup (bottom part) is composed of a pair of Si harmonic separators and controls the delay (Δt) between the two replicas of the reflected pulse. The two replica pulses with Δt delay time are focused by a concave mirror (SiC bulk or Sc/Si multilayer) onto a molecular beam of N_2. The fragment ions from N_2 molecules, which are introduced through a skimmer from a pulsed gas jet, are detected by a time-of-flight ion mass spectrograph, which is constructed from three electrodes, a flight tube, and a microchannel plate (MCP). Reproduced from [73] with permission from Nature.

the entrance and exit surfaces for maintaining the vacuum. The confocal parameters of the two beams are adjusted to be the same, matching the Gouy phase shift through the focus. The Xe target gas was statically filled into the interaction cell of 12 cm length. The harmonic signals were observed with a flat-field normal-incidence XUV spectrometer with a microchannel plate. A charge-coupled device camera detected the two-dimensional fluorescence from a phosphor screen placed behind the microchannel plate.

To evaluate the harmonic output energy, they first directly measured the total pulse energy of single-color-induced high-order harmonic generation (SC-HHG) from 11th to 21st orders by using an XUV photodiode. The values of spectral sensitivity and quantum efficiency of the XUV photodiode were obtained from [75] and were calibrated with 266-nm Q-switched YAG laser pulses in this experiment. To perfectly block the pump pulse and lower harmonic components (<11th), they inserted a 200-nm-thick Al filter supported by a mesh in front of the photodiode. Although the filter transmission of

Al can be estimated from the absorption coefficient, the Al filter transmission degrades with time because of oxidation. Therefore, the filter transmission for individual harmonics was calibrated using an XUV spectrometer. From the spectral distribution measured by the XUV spectrometer, they divided the total pulse energy among the six harmonic components. Thus, for SC-HHG they obtain the relationship between the harmonic spectral intensity and the pulse energy. In the TC-HHG experiments, they computed the area under the measured continuum spectrum by integration within a certain energy range, to evaluate the pulse energy in the cutoff region.

Figure 1.8 shows typical single-shot harmonic spectra of Xe gas as function of photon energy. The focused intensity in the TC case (ITC) was $5 \times 10^{13}\,\mathrm{W\,cm^{-2}}$ with an intensity ratio of $\sim$0.15. The harmonic yield is optimized by adjusting the focusing point with respect to the gas cell and by varying the gas pressure. As the difference between the dispersions for the 800- and 1300-nm pulses in Xe is negligible, the phase-matching condition can be satisfied by

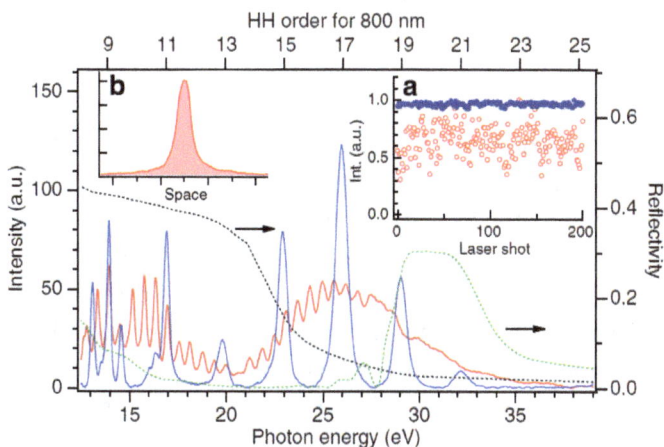

Fig. 1.8. Experimentally obtained single-shot harmonic spectra from Xe. Blue curve: SC laser field (800 nm: 11 mJ); red curve: TC laser field (800 nm: 9 mJ + 1300 nm: 2.5 mJ). Green dotted curve: reflectivity of the Sc/Si multilayer mirror; black dotted curve: reflectivity of the SiC bulk mirror. Inset: (a) measured intensity variation of the harmonic cutoff (29 eV: 19th order) over 200 laser shots. (b) The spatial profile of the cutoff harmonics of the TC-HHG. Reproduced from [73] with permission from Nature.

optimizing the gas pressure. The harmonic yield gradually increases as the Xe gas pressure increases, and exhibits a quadratic dependence on pressure. This means the conversion efficiency and the photon flux of HHG depend on the Xe gas pressure. The cutoff intensity of TC-HHG is optimized at a gas pressure of 0.8 Torr, which is in good agreement with the optimum gas pressure predicted by the phase-matching theory under these experimental conditions.

When the single-color field is focused to an intensity of 7×10^{13} W cm^{-2}, the harmonic spectrum has a discrete structure (blue curve). In the TC scheme (red curve), a quasi-continuous harmonic spectrum appears in the lower-order region, whereas the quasi-continuous harmonics disappear at harmonic orders higher than the 17th. One can clearly see a smooth continuum harmonic spectrum around the cutoff region. Note that the intensity of the TC-HHG is comparable to that of the SC-HHG. From the wavelength-scaling law of the TC-HHG under the 0.15 ratio, the conversion efficiency decreases only by a factor of $(1300/800)^{-0.7} = 0.71$ compared with the SC-HHG case. The maximum harmonic photon energy is slightly extended to 35 eV even though the focused intensity is lower than for the SC-HHG. This photon energy extension is explained by the cutoff formula for TC-HHG [65].

The inset (a) of Fig. 1.8 shows the measured intensity variation of the harmonic cutoff (29 eV; 19th order) over 200 laser shots. The intensity of the SC-HHG has a high stability, as shown by the blue points, whereas the intensity of the TC-HHG (red points) fluctuates between 1 and 0.3. These variations are due to the intensity changes of the TC field caused by random fluctuations in the CEP value of the non-CEP-stabilized Ti:sapphire laser and slight jitter of the relative phase introduced by the Mach–Zehnder interferometer used for creating the TC fields. According to the discussion of the variations of these two kinds of phase effects, the HHG yield at the cutoff should be maximized, when the continuous spectrum frequently appears for certain 'good' combinations of the CEP and relative phase, whereas only weak harmonics exhibiting the quasi-continuum spectrum is generated for 'bad' phase combinations. They have observed this correlation between the harmonic yield and the spectral shape in the experiment. Note that harmonic shots with the quasi-continuum spectrum hardly contribute to form the AC trace, because their intensity is quite low.

The output beam divergence of the TC-HHG at the cutoff was measured to be 0.5 mrad full-width half maximum with a Gaussian-like profile (see Fig. 1.8b inset). The almost perfect Gaussian profile of the harmonics suggests that there is no density disturbance due to high ionization in the harmonic medium. The pulse energy of the continuous spectrum (28–35 eV) is evaluated to be 1.3 μJ with a conversion efficiency of 1.1×10^{-4} from the input TC energy (800 nm: 9 mJ + 1300 nm: 2.5 mJ). This value is almost 100-fold higher than the highest pulse energies ever reported before [74]. The generation efficiency from the Ti:sapphire laser is 3.7×10^{-5}, because 30 mJ of Ti:sapphire energy was used for creating the TC field. Moreover, the midplateau region (14–29.5 eV) exhibiting a quasi-continuous spectrum in the TC-HHG has an output energy of 10 mJ.

To evaluate the temporal characteristics of the microjoule harmonic fields, they utilized an AC method using a spatially split autocorrelator. They employed two concave mirrors (R = 200 mm) for focusing the harmonics: one is a bulk SiC mirror for the mid-plateau and the other is a Sc/Si multilayer mirror for the cutoff. The reflectivity for each concave mirror is shown in Fig. 1.8. They have adopted the N^+ ion yield from fragmenting N_2 molecules as the indicator of the second-order AC signal. This choice is motivated by the findings of earlier experiment [76], in which the N^+ ion yield was sufficiently high to obtain the AC trace of an attosecond pulse in the relevant photon energy region.

The dotted curves in the top panels of Fig. 1.9a,b are the measured AC traces of the mid-plateau TC-HHG and SC-HHG fields, respectively. The signals at each delay time are obtained by accumulating signals for 1×10^3 laser shots. They can see distinct peaks emerge at every half period (1.33 fs) of the Ti:sapphire laser field in the AC trace depicted in the top panel of Fig. 1.9b.

This result ensures that the AC measurement can certainly reveal the temporal shape in the attosecond regime. In contrast, the four peaks corresponding to multiple bursts on both side of the main

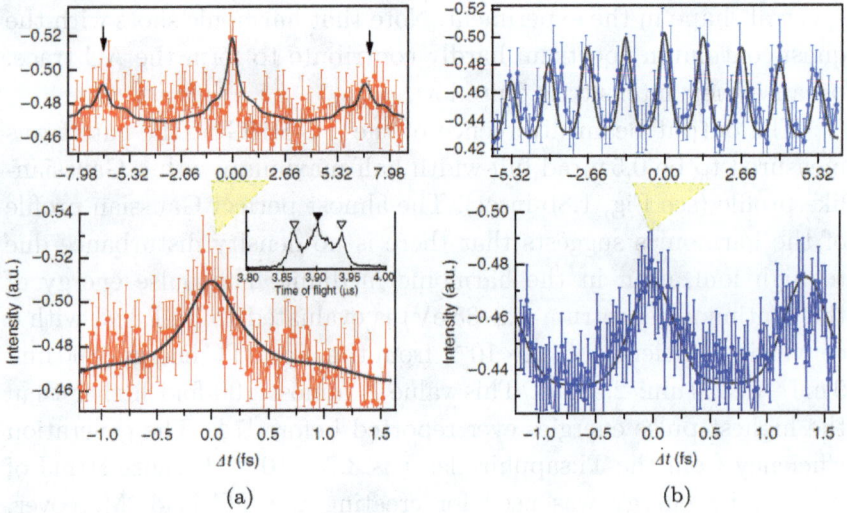

Fig. 1.9. Measured AC traces from the side peak of N^+ ion signals. (a) Mid-plateau of TC-HH; (b) mid-plateau of SC-HH. The time resolutions of the top and bottom panels correspond to 148 and 28 as, respectively. The error bars show the s.d. of each data point. The grey solid profiles are the AC traces of the simulated HH spectrum. Inset of (a): measured TOF spectrum of ions at around $m/z \approx 14$. The inverted triangles and black-filled inverted triangle correspond to the kinetic energies of 3 and 0 eV, respectively. Reproduced from [73] with permission from Nature.

peak (0 fs) are well suppressed in the TC-HH, even though we can observe the side peaks at ±6.7 fs (indicated by arrows in Fig. 1.9a) with a relative AC amplitude of 0.45 compared with the main peak. These two side peaks are consistent with our simulated result of the TC-HHG.

To observe the pulse duration more precisely, they increased the time resolution of the AC measurement by reducing the delay step from 148 to 28 as. The bottom panels in Fig. 1.9 show the measured AC signal for SC-HHG (blue points) and TC-HHG (red points). In the mid-plateau of TC-HHG, the full-width half maximum of the AC was measured to be 530 ± 75 as by fitting the trace with a Gaussian function (not shown). The pulse duration is thus 375 as. On the other hand, the measured AC duration for the SC-HHG case was 490 ± 60 as. Thus, the duration of each peak in the attosecond pulse train is evaluated to be 350 as, which is slightly shorter than the pulse duration of the TC-HHG. The broadening of the AC duration of the TC-HHG compared with the SC-HHG case is due to the fact that the AC trace of the TC-HHG originates from the temporal profiles averaged over the random CEP and slightly fluctuating relative phase. In addition, the harmonic spectral shapes are different in both cases.

Since the first demonstration of an IAP was reported in 2001 [8], continuous effort to increase the IAP energy [72] has been made by many research groups around the world over the past decade, as illustrated in Fig. 1.10. Unfortunately, the output energy of IAPs has not exceeded the nanojoule level thus far. Owing to the low photon flux, the temporal characterization of the IAP was generally achieved using the attosecond streaking method employing frequency-resolved optical gating for complete reconstruction of attosecond bursts (FROG-CRAB). The authors of the reviewed paper have improved the available IAP energy for nonlinear optics

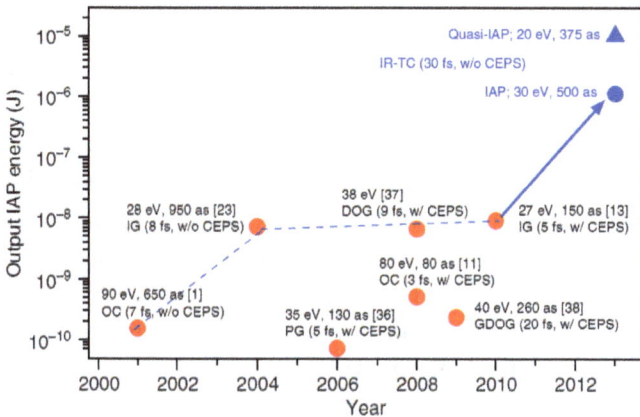

Fig. 1.10. Progress of the energy of IAPs used in attosecond experiments. Red circles: previous milestone results using various schemes; blue points: our HHG source. IG, ionization gating; PG, polarization gating; DOG, double optical gating; GDOG, generalized DOG. We indicate the required laser pulse duration and CEP stabilization (CEPS) for each generation scheme. Note the dramatic leap in IAP energy, which is used in an attosecond experiment, by two orders of magnitude achieved in the reviewed work. Reproduced from [73] with permission from Nature.

experiments by two orders of magnitude compared with the highest output energy reported previously [54]. Their method features good energy scalability and high conversion efficiency owing to the phase-matching technique. They directly characterized the temporal profile of our IAP by an AC method using the nonlinear interaction of N_2. In fact, the AC traces clearly demonstrate that the discussed HH source is capable of not only inducing nonlinear phenomena but also having attosecond temporal resolution. This experiment also demonstrates the capability of the proposed technique to observe nonlinear attosecond dynamics in real time with IAPs by time-resolved spectroscopy employing a non-sequential two-XUV-photon process in a molecule. Moreover, the obtained intense IAPs are useful as the external injector for a seeded FEL [77] in the XUV region.

1.2.3. *Generation of coherent radiation in the water window region at 1 kHz repetition rate using a mid-infrared pump source*

There has long been a quest for a table-top coherent x-ray source in the water window (4.37–2.33 nm) region by which three-dimensional imaging of living cells can be achieved with both high contrast and high spatial resolution [78]. So far, the coherent water window radiation has been demonstrated by several groups using high-order harmonic generation in noble gas media with 800 nm pump laser sources [79–82]. However, when an 800 nm pump laser is used for generating the coherent x-ray in water window, high peak intensity near or even beyond the ionization saturation intensity must be used to provide sufficiently high kinetic energies to those electrons, which, after being tunnel ionized in the laser field, are able to recombine with their parent ions and release their energies by emitting XUV photons. The strong ionization in turn leads to low conversion efficiency owing to the spoiled phasematching conditions. Fortunately, the well-known three-step model of HHG predicts that the cutoff law obeys $E_{\text{cutoff}} = I_p + 3.17 U_p$ (here I_p is the ionization potential and U_p is the ponderomotive energy) [83–85], implying that instead of increasing the peak intensity, one can also extend the cutoff energy of HHG by increasing the driving wavelength [11,65,86,87]. Recently, a coherent water window x-ray has been demonstrated by HHG with a gas jet using a homebuilt high-energy MIR pump laser operating at 10 Hz repetition rate and 1550 nm wavelength [45,65].

In [88], the authors have demonstrated the generation of the water window x-ray at 1 kHz repetition rate in a 4-mm-long gas cell using a commercially available mid-IR pump laser. In addition, optimization of the HHG signal in the water window region was achieved by tuning the gas pressure as well as the focal position. Their scheme was similar to one described in previous section. In their experiment, wavelength-tunable MIR laser pulses were generated by an optical parametric amplifier (HE-TOPAS, Light Conversion, Inc.) pumped by a commercial Ti:sapphire laser system (Legend Elite-Duo, Coherent, Inc.). The Ti:sapphire laser, operated at a repetition rate of 1 kHz, provided 40 fs (FWHM) laser pulses with a central wavelength at 795 nm and a single pulse energy of 6.4 mJ. At the 1500 nm wavelength, the maximum pulse energy that the OPA can offer was 1.6 mJ, and the pulse duration was 50 fs. To separate the signal and idler beams, dielectric coated broadband mirrors with high reflection coatings were used. A fused-silica lens with a focal length of 15 cm at 1500 nm wavelength was used to focus the IR beams, which can be translated along the beam propagation direction by a high-precision translation stage. A 4-mm-long gas cell filled with neon gas was mounted inside the vacuum chamber for generating high-order harmonics, and variable gas pressures in a range from 0 to 125 mbar were used in this experiment. A flat-field grating spectrometer equipped with a soft-x-ray CCD camera was used for measurement of the HHG spectra. The spectrometer consisted of a spherical concave gold mirror and a flat-field grating. To block the low-order harmonics and the residual IR driving pulses, a 150-nm-thick aluminum foil was placed at the entrance of the spectrometer.

When the 1500 nm pump pulse is focused into the gas cell, high-order harmonics in the water window region can be easily produced. A typical HHG spectrum showing a cutoff beyond 400 eV is presented by the solid curve in Fig. 1.11. The dip around the carbon K edge (284 eV) should be caused by a thin layer of pump oil absorbed on the

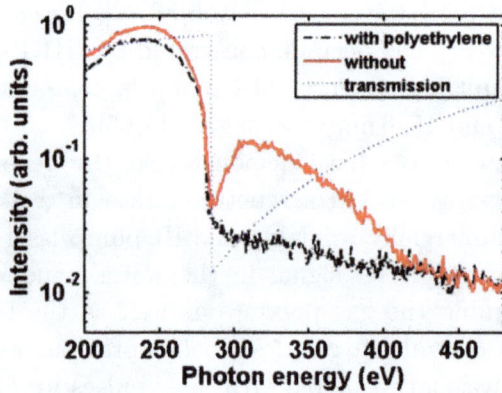

Fig. 1.11. Normalized HHG spectra obtained in neon with (dashed-dotted curve) and without (solid curve) two 500-nm-thick polyethylene films. Dotted curve, transmission curve of the polyethylene film. Reproduced from [88] with permission from Optical Society of America.

optical components of the spectrometer or on the soft-x-ray CCD, as had been pointed out in [89]. The highest photon energy detected was 420 eV. It is noteworthy that the grating (1200 grooves/mm) they employed was optimized to cover the 50–200 eV photon energy range, which led to a rapid drop in efficiency when the photon energy was higher than 300 eV. To confirm the generation of the water window x ray, HHG spectrum is recorded with two additional 500-nm-thick polypropylene films placed before the aluminum filter, as shown by the dotted-dashed curve in Fig. 1.11. In this case, the high-order harmonics beyond the carbon K edge completely disappeared in the HHG spectrum.

The authors of the refereed paper then optimized the HHG in the water window region by tuning the focal position and gas pressure. For these measurements, the peak intensity of the mid-IR driving laser was set at $4.7 \times 10^{14} \, W \, cm^{-2}$. HHG spectra measured at a fixed pressure of 61 mbar but variable position of focal spot of the driving pulses are shown in Fig. 1.12(a). When the focal spot was located 3.5 mm after the center of the gas cell (corresponding to a focal position of +3.5 mm), there was no HHG flux in the water window region, because the focal spot of driving laser was out of the gas cell and the laser intensity in the interaction area was low. When

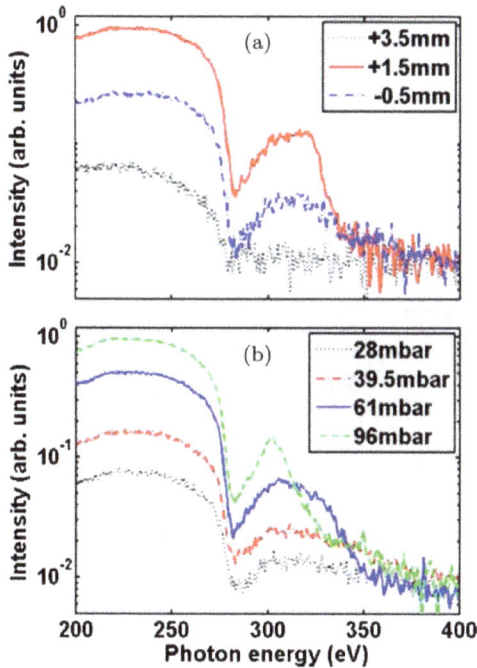

Fig. 1.12. Normalized HHG spectra captured (a) at different focal spot positions and (b) under different neon pressures. The intensity of the 1500 nm driving laser is fixed at 4.7×10^{14} W cm^{-2}. Reproduced from [88] with permission from Optical Society of America.

the focal spot was shifted to 1.5 mm after the center of the gas cell, the HHG intensity significantly increases, and a sharp cutoff around 330 eV can be observed. When the focal spot is further shifted to 0.5 mm before the center of the gas cell (corresponding to a focal position of 0.5 mm), the HHG intensity decreases again, while the cutoff remains almost the same. Generally, the cutoff and intensity of HHG are determined by both the intensity of driving laser and the phase-matching condition. Thus, there was usually no simple relationship between the intensity and cutoff of a HHG spectrum.

Figure 1.12(b) shows the HHG spectra captured under different gas pressures when the focal position was fixed at +1.5 mm (i.e., 1.5 mm after the center of the gas cell). Clearly, one can see that when the gas pressure increased, the cutoff decreased while the efficiency of HHG increased. This could be understood by the fact that by increasing the gas pressure, there will be more atoms participating in the HHG process, resulting in an increase in the conversion efficiency. However, high gas pressure will also lead to high density of free electrons, which can strongly spoil the phase-matching condition [90]. This is particularly true for high-energy photons for which the optimal phase matching is more difficult to achieve, so that the cutoff energy decreases with increasing gas pressure.

Finally, they compared the HHG spectra obtained at 800, 1300, and 1500 nm wavelengths in neon gas at a nearly constant focal intensity of $6.0 \times 10^{14}\,\mathrm{W\,cm^{-2}}$ of the pump laser beams. In all measurements at the three different wavelengths, they individually optimized the HHG yield as well as the cutoff by tuning the gas pressure and focal position. The results are shown in Fig. 1.13. With the 800 nm driving pulse, the highest cutoff energy achieved was only 135 eV. And with the 1300 nm driving pulse, the harmonic spectrum in the water window can be merely observed. In the case of the 1500 nm driving pulse, the cutoff is greatly extended to 420 eV. Therefore, the cutoff energy roughly obeys the wavelength scaling law of HHG (e.g., $E_{\mathrm{cutoff}} \propto \lambda^2$) [10]. Furthermore, the second-order diffraction spectrum was found in Fig. 1.13 because of the low transmission efficiency of the aluminum filter in the spectral range of 75–150 eV. Note that the second-order diffraction spectrum does not exactly correspond to a half of the photon energy of the first-order diffraction spectrum because of the varying line spacing of the grating. Besides the high-order diffraction, other factors such as spectral dependence of the HHG detection system and some unidentified single atom and collective effects [91] may also contribute to the modulation of the spectra in Fig. 1.13.

1.2.4. *Various approaches in the HHG using MIR pulses*

There are different other ways to generate HHG using MIR pulses. A two-color synthesis scheme to achieve optimal harmonic yields ranging from the extreme ultraviolet to soft X-rays was proposed

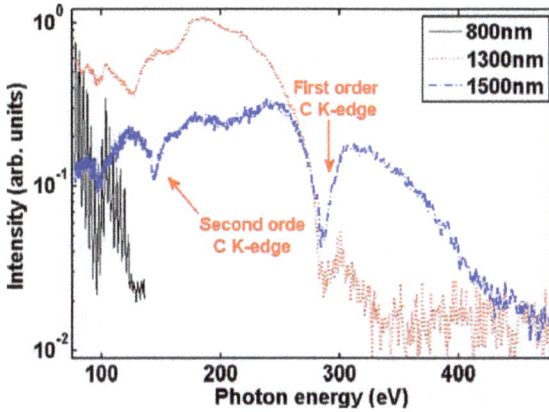

Fig. 1.13. Normalized HHG spectra optimized at different wavelengths of the driving pulses. Solid curve, 800 nm; dotted curve, 1300 nm; dashed-dotted curve, 1500 nm. The intensity of the laser beams is fixed at $6.0 \times 10^{14}\,\mathrm{W\,cm^{-2}}$ for the three wavelengths. Reproduced from [88] with permission from Optical Society of America.

in [92]. Between the two colors, one is the strong fundamental MIR radiation, the other is its weak third harmonic. For targets of argon, neon and helium, they optimized the shortest wavelength of the fundamental pulse that should be used for a given cutoff energy. Their results also showed that optimized two-color field can always enhance the HHG yield, when compared with the single-color one under the same generating condition, and requiring only about 5% intensity for the third harmonic, which is easily reachable. Some progress in waveform synthesis and optimization has been made in previous studies [93–95]. Still much remains to be explored, including waveform control with multi-color lasers, or full phase control of a broadband supercontinuum over two octaves. With the emergence of hundreds kHz and MHz MIR lasers, the simple two-color optimization, together with favorable phase matching conditions, leads us to believe that high-order harmonics will soon be ready to become useful coherent tabletop light sources from the extreme ultraviolet to the x-rays.

In the time domain, they also serve as intense attosecond pulses that can be used for attosecond-pump and attosecond-probe experiments for studying dynamics of electrons in matter in their intrinsic timescales. High harmonics extending to x-rays have been generated from gases by intense lasers. To establish these coherent broadband radiations as an all-purpose tabletop light source for general applications in science and technology, new methods are needed to overcome the present low conversion efficiencies. The authors of [92] have shown that the conversion efficiency may be drastically increased with an optimized two-color pulse. By employing an optimally synthesized 2-μm MIR radiation and a small amount of its third harmonic, they have shown that harmonic yields from sub- to few-keV energy can be increased typically by an order of magnitude over the optimized SC one. By combining with favorable phase-matching and together with the emerging high-repetition MHz mid-infrared lasers, they anticipate that the efficiency of harmonic yield can be increased by four to five orders in the near future, thus paving the way for employing high harmonics as useful broadband tabletop light sources from the extreme ultraviolet to the x-rays, as well as providing new tools for interrogating ultrafast dynamics of matter at attosecond timescales.

Generation of bright isolated attosecond soft x-ray pulses driven by multicycle MIR lasers has been analyzed in [96]. HHG driven by femtosecond lasers makes it possible to capture the fastest dynamics in molecules and materials. However, to date the shortest subfemtosecond pulses have been produced only in the extreme UV region of the spectrum below 100 eV, which limits the range of materials and molecular systems that can be explored. The authors have experimentally demonstrated a remarkable convergence of physics: when MIR lasers are used to drive HHG, the conditions for optimal bright, soft x-ray generation naturally coincide with the generation of isolated attosecond pulses. The temporal window over which phase matching occurs shrinks rapidly with increasing driving laser wavelength, to the extent that bright isolated attosecond pulses are the norm for 2-μm driving lasers. Harnessing this realization, they have experimentally demonstrated the generation of isolated soft x-ray attosecond pulses at photon energies up to 180 eV, with a transform limit of 35 attoseconds. Most surprisingly, advanced theory showed that in contrast with attosecond pulse generation in the XUV, long duration, 10-cycle, driving laser pulses are required to generate isolated soft x-ray bursts efficiently, to mitigate group velocity walk-off between the laser and the x-ray fields that otherwise limit the conversion efficiency. This work demonstrates a clear and straightforward approach for robustly generating bright isolated attosecond pulses of electromagnetic radiation throughout the soft x-ray region of the spectrum.

A typical multicycle driver pulse will produce a train of attosecond pulses (AP). The key principle to decrease the number of AP in the train is to reduce the periodicity of the HHG process until the emission is confined to a very short time interval. In the frequency domain, this amounts to the generation of a spectral supercontinuum [97], which can be achieved by reducing the number of cycles of the driver [8] and/or by introducing a gate or switch of subcycle duration to longer pulses. As already mentioned, the shortest single AP produced so far, lasting less than 100 as, have been generated from 3.3-fs pulses [64]. These approaches also rely on spectral filtering of AP to select emission from only the highest intensity part of the pulse, as well as ionization gating to terminate the process.

However, few-cycle pulses are technically challenging to produce [98, 99]. Therefore, other approaches are being pursued to relax the need to produce ultrashort driving pulses. For example, polarization gating (PG) in combination with few-cycle pulses, which has recently yielded 130-as pulses [55], exploits the high sensitivity of the recollision step to the field's ellipticity [100] by rapidly sweeping the polarization state of the field from circular to linear within one optical cycle [101–103]. Another technique has been demonstrated in which the addition of a phase-locked second harmonic field [104–107] inhibits the ionization on one polarity of the optical field and causes the HHG spectrum to show a mixture of even and odd harmonics. This results in an AP train with a separation of a full optical period. Combined with PG in a technique called double optical gating, an XUV supercontinuum capable of supporting 130-as pulses was generated using 9-fs driving pulses [59]. Other approaches of combining different frequencies have been studied theoretically

[66, 108, 109]. A recent experiment demonstrated increased spectral density when two incommensurate laser frequencies were combined [91], creating the conditions necessary for temporal confinement of attosecond pulse generation.

In [110], the feasibility of a new gating technique was experimentally demonstrated, which relies on the mixing of two or three multicycle laser pulses at incommensurate frequencies to further modify the subcycle shape of the electric field in order to produce single AP. The mixing of incommensurate driving frequencies for the purpose of synthesizing single AP has only been studied theoretically [111] but has never been realized in practice until now. Previous experiments used this approach to achieve tunability of HHG [112, 113]. Those results clearly show that the addition of a second (and then a third) optical field whose frequency is not commensurate with the primary field can create additional frequency components in the HHG spectrum. Depending on the frequency ratio between the fields, this can result in a quasicontinuous spectrum in the plateau region. They also showed numerically that this subcycle field control will lead to the possibility of generating a single attosecond pulse from a combination of three 24-fs laser pulses, similar to a proposal by [114] to combine three regions of a broadband spectrum to produce a single-cycle pulse.

1.3. Conclusions to Chapter 1

We presented different schemes of MIR sources. In particular, we have discussed in detail a simple scheme to generate 10 mJ, 30 fs pulses centered around 1.8 μm at a repetition rate of 100 Hz corresponding to 0.33 TW peak power [50]. This source is very well suited for driving a large variety of nonlinear optical processes opening many scientific opportunities such as synchronized high-field THz pump pulses with ultrashort soft X-ray pulses for probing ultrafast phenomena in the condensed matter. Moreover, using the concept of sub-cycle phase matching, this laser should be ideal to generate isolated attosecond pulses, but remains to be confirmed using the well established streaking technique.

Those and similar sources of MIR radiation were used for the HHG and attosecond pulses generation. We have analyzed some of those results. We have discussed the generation of a continuum high-order harmonic spectrum by mixing multicycle two-color laser fields. By mixing 30 fs, 800 nm and 40 fs, 1300 nm pulses, the authors of [65] obtained a continuum harmonic spectrum around the cutoff region without CEP stabilization, although the intensity of the cutoff HH is affected by the CEP drift. It should be emphasized that this TC scheme enables us to markedly suppress the ionization probability compared with those in the OC, polarization gating, DOG, and GDOG methods. This is a major advantage for efficiently generating intense IAPs from neutral-medium conditions, because neutral media enable us not only to apply the most appropriate phase-matching technique but also to apply an energy-scaling scheme. This new scheme greatly reduces the requirements for the pump laser system used for generating an IAP, and it allows us to use a well-established conventional high-power pumping laser. This method has the potential to greatly increase the intensity of IAPs by using an energy-scaling method.

High-energy isolated attosecond pulses required for the most intriguing nonlinear attosecond experiments as well as for attosecond-pump/attosecond-probe spectroscopy are still lacking at present. In the reviewed paper [73], a robust generation method of intense isolated attosecond pulses, which enable a nonlinear attosecond optics experiment to be performed, was demonstrated. By combining a two-color field synthesis and an energy-scaling method of high-order harmonic generation, the maximum pulse energy of the isolated attosecond pulse reaches as high as 1.3 μJ. The generated pulse with a duration of 500 as, as characterized by a nonlinear autocorrelation measurement, is the shortest and highest-energy pulse ever with the ability to induce nonlinear phenomena. The peak power of our tabletop light source reaches 2.6 GW, which even surpasses that of an XUV free-electron laser.

Another important component of these studies is the extension of the photon energy of generated XUV pulses. In [88], the generation of the coherent water window x-ray using a wavelength tunable mid-IR pump laser was reported. Significant enhancement of the water window x-ray can be achieved by tuning the gas pressure and focal position, indicating the importance of phase matching. The table-top 1 kHz repetition-rate water window x-ray source driven by mid-IR pump source opens up possibilities for applications such as living cell imaging, nanolithography, and so on.

These and numerous other studies have demonstrated the usefulness in application of the gases for the HHG using MIR pulses. One can anticipate further developments of this technique using another nonlinear optical medium, laser-produced plasma. In the following chapters, we will show the advantages of plasma media for the HHG using coherent MIR radiation.

References

[1] F. Krausz and M. Ivanov, *Rev. Mod. Phys.* **81**, 163 (2009).
[2] S. Woutersen, U. Emmerichs, and H. J. Bakker, *Science* **278**, 658 (1997).
[3] F. K. Tittel, D. Richter, and A. Fried, *Top. Appl. Phys.* **89**, 458 (2003).
[4] B. Jean and T. Bende, *Top. Appl. Phys.* **89**, 530 (2003).
[5] J. Reichert, R. Holzwarth, T. Udem, and T. W. Hänsch, *Opt. Commun.* **172**, 59 (1999).
[6] S. A. Diddams, *J. Opt. Soc. Am. B* **27**, B51 (2010).
[7] S. Schliesser, N. Picqué, and T. W. Hänsch, *Nat. Photonics* **6**, 440 (2012).
[8] M. Hentschel, R. Kienberger, C. Spielmann, G. A. Reider, N. Milosevic, T. Brabec, P. Corkum, U. Heinzmann, M. Drescher, and F. Krausz, *Nature* **414**, 509 (2001).
[9] J. Biegert, P. K. Bates, and O. Chalus, *IEEE J. Sel. Top. Quantum Electron. Ultrafast Sci. Technol.* **18**, 531 (2012).
[10] J. Tate, T. Auguste, H. G. Muller, P. Salières, P. Agostini, and L. F. DiMauro, *Phys. Rev. Lett.* **98**, 013901 (2007).
[11] T. Popmintchev, M. Chen, O. Cohen, M. Grisham, J. Rocca, M. Murnane, and H. Kapteyn, *Opt. Lett.* **33**, 2128 (2008).
[12] M. V. Frolov, N. L. Manakov, and A. F. Starace, *Phys. Rev. Lett.* **100**, 173001 (2008).
[13] S. L. Cousin, F. Silva, S. Teichmann, M. Hemmer, B. Buades, and J. Biegert, *Opt. Lett.* **39**, 5383 (2014).
[14] F. Silva, S. Teichmann, S. L. Cousin, M. Hemmer, and J. Biegert, *Nat. Commun.* **6**, 6611 (2015).
[15] F. Silva, P. K. Bates, A. Esteban-Martin, M. Ebrahim-Zadeh, and J. Biegert, *Opt. Lett.* **37**, 933 (2012).

[16] P. Whalen, P. Panagiotopoulos, M. Kolesik, and J. V. Moloney, *Phys. Rev. A* **89**, 023850 (2014).

[17] T. Auguste, P. Salières, A. S. Wyatt, A. Monmayrant, I. A. Walmsley, E. Cormier, A. Zaïr, M. Holler, A. Guandalini, F. Schapper, J. Biegert, L. Gallmann, and U. Keller, *Phys. Rev. A* **80**, 033817 (2009).

[18] H. R. Reiss, *Phys. Rev. Lett.* **101**, 043002 (2008).

[19] I. Jovanovic, G. Xy, and S. Wandel, *Phys. Proc.* **52**, 68 (2014).

[20] D. Sanchez, M. Hemmer, M. Baudisch, S. L. Cousin, K. Zawilski, P. Schunemann, O. Chalus, C. Simon-Boisson, and J. Biegert, *Optica* **3**, 147 (2016).

[21] G. Andriukaitis, T. Balčiūnas, S. Ališauskas, A. Pugžlys, A. Baltuška, T. Popmintchev, M.-C. Chen, M. M. Murnane, and H. C. Kapteyn, *Opt. Lett.* **36**, 2755 (2011).

[22] T. Popmintchev, M.-C. Chen, D. Popmintchev, P. Arpin, S. Brown, S. Ališauskas, G. Andriukaitis, T. Balčiūnas, O. D. Mücke, A. Pugzlys, A. Baltuška, B. Shim, S. E. Schrauth, A. Gaeta, C. Hernandez-Garcia, L. Plaja, A. Becker, A. Jaron-Becker, M. M. Murnane, and H. C. Kapteyn, *Science* **336**, 1287 (2012).

[23] D. Kartashov, S. Ališauskas, G. Andriukaitis, A. Pugžlys, M. Shneider, A. Zheltikov, S. L. Chin, and A. Baltuška, *Phys. Rev. A* **86**, 033831 (2012).

[24] D. Kartashov, S. Ališauskas, A. Pugžlys, A. A. Voronin, A. M. Zheltikov, and A. Baltuška, *Opt. Lett.* **37**, 2268 (2012).

[25] E. E. Serebryannikov and A. M. Zheltikov, *Phys. Rev. Lett.* **113**, 043901 (2014).

[26] A. A. Lanin, A. A. Voronin, E. A. Stepanov, A. B. Fedotov, and A. M. Zheltikov, *Opt. Lett.* **39**, 6430 (2014).

[27] Y. Nomura, H. Shirai, K. Ishii, N. Tsurumachi, A. A. Voronin, A. M. Zheltikov, and T. Fuji, *Opt. Express* **20**, 24741 (2012).

[28] D. Brida, M. Marangoni, C. Manzoni, S. De Silvestri, and G. Cerullo, *Opt. Lett.* **33**, 2901 (2008).

[29] F. Silva, D. R. Austin, A. Thai, M. Baudisch, M. Hemmer, D. Faccio, A. Couairon, and J. Biegert, *Nat. Commun.* **3**, 807 (2012).

[30] M. Hemmer, M. Baudisch, A. Thai, A. Couairon, and J. Biegert, *Opt. Express* **21**, 28095 (2013).

[31] A. A. Voronin and A. M. Zheltikov, *Phys. Rev. A* **90**, 043807 (2014).

[32] C. C. Wang and G. W. Racette, *Appl. Phys. Lett.* **6**, 169 (1965).

[33] S. E. Harris, M. K. Oshman, and R. L. Byer, *Phys. Rev. Lett.* **18**, 732 (1967).

[34] V. V. Yakovlev, B. Kohler, and K. R. Wilson, *Opt. Lett.* **19**, 2000 (1994).

[35] F. Rotermund, V. Petrov, and F. Noack, *Opt. Commun.* **185**, 177 (2000).

[36] M. K. Reed, M. K. Steiner-Shepard, and D. K. Negus, *Opt. Lett.* **19**, 1855 (1994).

[37] C. I. Blaga, J. Xu, A. D. DiChiara, E. Sistrunk, K. Zhang, P. Agostini, T. A. Miller, L. F. DiMauro, and C. D. Lin, *Nature* **483**, 194 (2012).

[38] M. Clerici, M. Peccianti, B. E. Schmidt, L. Caspani, M. Shalaby, M. Giguère, A. Lotti, A. Couairon, F. Légaré, T. Ozaki, D. Faccio, and R. Morandotti, *Phys. Rev. Lett.* **110**, 253901 (2013).

[39] J. Weisshaupt, V. Juve, M. Holtz, S. Ku, M. Woerner, T. Elsaesser, S. Ališauskas, A. Pugžlys, and A. Baltuška, *Nat. Photonics* **8**, 927 (2014).

[40] A. Dubietis, G. Jonušauskas, and A. Piskarskas, *Opt. Commun.* **88**, 437 (1992).

[41] D. Brida, C. Manzoni, G. Cirmi, M. Marangoni, S. Bonora, P. Villoresi, S. De Silvestri, and G. Cerullo, *J. Opt.* **12**, 013001 (2010).

[42] C. Vozzi, F. Calegari, E. Benedetti, S. Gasilov, G. Sansone, G. Cerullo, M. Nisoli, S. De Silvestri, and S. Stagira, *Opt. Lett.* **32**, 2957 (2007).

[43] M. Giguère, B. E. Schmidt, A. D. Shiner, M.-A. Houle, H.-C. Bandulet, G. Tempea, D. M. Villeneuve, J.-C. Kieffer, and F. Légaré, *Opt. Lett.* **34**, 1894 (2009).

[44] B. E. Schmidt, P. Bèjot, M. Giguère, A. D. Shiner, C. Trallero-Herrero, E. Bisson, J. Kasparian, J.-P. Wolf, D. M. Villeneuve, J.-C. Kieffer, P. B. Corkum, and F. Légaré, *Appl. Phys. Lett.* **96**, 121109 (2010).

[45] E. J. Takahashi, T. Kanai, Y. Nabekawa, and K. Midorikawa, *Appl. Phys. Lett.* **93**, 041111 (2008).

[46] O. Chalus, P. K. Bates, M. Smolarski, and J. Biegert, *Opt. Express* **17**, 3587 (2009).

[47] H. Fattahi, H. G. Barros, M. Gorjan, T. Nubbemeyer, B. Alsaif, C. Y. Teisset, M. Schultze, S. Prinz, M. Haefner, M. Ueffing, A. Alismail, L. Vamos, A. Schwarz, O. Pronin, J. Brons, X. T. Geng, G. Arisholm, M. Ciappina, V. S. Yakovlev, D.-E. Kim, A. M. Azzeer, N. Karpowicz, D. Sutter, Z. Major, T. Metzger, and F. Krausz, *Optica* **1**, 45 (2014).

[48] B. E. Schmidt, N. Thiré, M. Boivin, A. Laramée, F. Poitras, G. Lebrun, T. Ozaki, H. Ibrahim, and F. Légaré, *Nat. Commun.* **5**, 3643 (2014).

[49] A. Vaupel, N. Bodnar, B. Webb, L. Shah, and M. Richardson, *Opt. Eng.* **53**, 051507 (2013).

[50] N. Thiré, S. Beaulieu, V. Cardin, A. Laramée, V. Wanie, B. E. Schmidt, and F. Légaré, *Appl. Phys. Lett.* **106**, 091110 (2015).

[51] S. Klingebiel, I. Ahmad, C. Wandt, C. Skrobol, S. A. Trushin, Z. Major, F. Krausz, and S. Karsch, *Opt. Express* **20**, 3443 (2012).

[52] A. Baltuška, T. Udem, M. Uiberacker, M. Hentschel, E. Goulielmakis, C. Gohle, R. Holzwarth, V. S. Yakovlev, A. Scrinzi, T. W. Hänsch, and F. Krausz, *Nature (London)* **421**, 611 (2003).

[53] R. Kienberger, E. Goulielmakis, M. Uiberacker, A. Basltuska, V. Yakovlev, F. Bammer, A. Scrinzi, T. Westerwalbesloh, U. Kleinberger, U. Heinzmann, M. Drescher, and F. Krausz, *Nature (London)* **427**, 817 (2004).

[54] T. Sekikawa, A. Kosuge, T. Kanai, and S. Watanabe, *Nature (London)* **432**, 605 (2004).

[55] G. Sansone, E. Benedetti, F. Calegari, C. Vozzi, L. Avaldi, R. Flammini, L. Poletto, P. Villoresi, C. Altucci, R. Velotta, S. Stagira, S. De Silvestri, and M. Nisoli, *Science* **314**, 443 (2006).

[56] P. Tzallas, D. Charalambidis, N. A. Papadogiannis, K. Witte, and G. D. Tsakiris, *Nature* **426**, 267 (2003).

[57] Y. Nabekawa, T. Shimizu, T. Okino, K. Furusawa, H. Hasegawa, K. Yamanouchi, and K. Midorikawa, *Phys. Rev. Lett.* **96**, 083901 (2006).

[58] E. Takahashi, Y. Nabekawa, T. Otsuka, M. Obara, and K. Midorikawa, *Phys. Rev. A* **66**, 021802 (2002).

[59] H. Mashiko, S. Gilbertson, C. Li, S. D. Khan, M. M. Shakya, E. Moon, and Z. Chang, *Phys. Rev. Lett.* **100**, 103906 (2008).

[60] Y. Oishi, M. Kaku, A. Suda, F. Kannari, and K. Midorikawa, *Opt. Express* **14**, 7230 (2006).

[61] Y. Zheng. Z. Zeng, X. Li, X. Chen, P. Liu, H. Xiong, H. Lu, S. Zhao, P. Wei, L. Zhang, Z. Wang, J. Liu, Y. Cheng, R. Li, and Z. Xu, *Opt. Lett.* **33**, 234 (2008).

[62] P. Tzallas, E. Skantzakis, C. Kalpouzos, E. P. Benis, G. D. Tsakiris, and D. Charalambidis, *Nature Phys.* **3**, 846 (2007).

[63] Q. Zhang, P. Lu, P. Lan, W. Hong, and Z. Yang, *Opt. Express* **16**, 9795 (2008).

[64] E. Goulielmakis, M. Schultze, M. Hofstetter, V. S. Yakovlev, J. Gagnon, M. Uiberacker, A. L. Aquila, E. M. Gullikson, D. T. Attwood, R. Kienberger, F. Krausz, and U. Kleineberg, *Science* **320**, 1614 (2008).

[65] E. J. Takahashi, P. Lan, O. D. Mücke, Y. Nabekawa, and K. Midorikawa, *Phys. Rev. Lett.* **104**, 233901 (2010).

[66] T. Pfeifer, L. Gallmann, M. J. Abel, P. M. Nagel, D. M. Neumark, and S. R. Leone, *Phys. Rev. Lett.* **97**, 163901 (2006).

[67] B. Kim, J. Ahn, Y. Yu, Y. Cheng, Z. Xu, and D. E. Kim, *Opt. Express* **16**, 10 331 (2008).

[68] E. J. Takahashi, T. Kanai, K. L. Ishikawa, Y. Nabekawa, and K. Midorikawa, *Phys. Rev. Lett.* **101**, 253901 (2008).

[69] A. Baltuška T. Fuji, and T. Kobayashi, *Phys. Rev. Lett.* **88**, 133901 (2002).

[70] T. Morishita, A.-T. Le, Z. Chen, and C. D. Lin, *Phys. Rev. Lett.* **100**, 013903 (2008).

[71] M. D. Perry and J. K. Crane, *Phys. Rev. A* **48**, R4051 (1993).

[72] G. Sansone, L. Poletto, and M. Nisoli, *Nat. Photonics* **5**, 655 (2011).

[73] E. J. Takahashi, P. Lan, O. D. Mücke, Y. Nabekawa, and K. Midorikawa, *Nat. Commun.* **4**, 2691 (2013).

[74] F. Ferrari, F. F. Calegari, M. Lucchini, C. Vozzi, S. Stagira, G. Sansone, and M. Nisoli, *Nat. Photonics* **4**, 875 (2010).

[75] M. Krumrey, E. Tegeler, R. Goebel, and R. Kohler, *Rev. Sci. Instrum.* **66**, 4736 (1995).

[76] Y. Nabekawa, T. Shimizu, T. Okino, K. Furusawa, H. Hasegawa, K. Yamanouchi, and K. Midorikawa, *Phys. Rev. Lett.* **97**, 153904 (2006).

[77] T. Togashi, E. J. Takahashi, K. Midorikawa, M. Aoyama, K. Yamakawa, T. Sato, A. Iwasaki, S. Owada, T. Okino, K. Yamanouchi, F. Kannari, A. Yagishita, H. Nakano, M. E. Couprie, K. Fukami, T. Hatsui, T. Hara, T. Kameshima, H. Kitamura, N. Kumagai, S. Matsubara, M. Nagasono,

H. Ohashi, T. Ohshima, Y. Otake, T. Shintake, K. Tamasaku, H. Tanaka, T. Tanaka, K. Togawa, H. Tomizawa, T. Watanabe, M. Yabashi, and T. Ishikawa, *Opt. Express* **19**, 317 (2011).

[78] J. C. Solem and G. C. Baldwin, *Science* **218**, 229 (1982).

[79] C. Spielmann, N. H. Burnett, S. Sartania, R. Koppitsch, M. Schnürer, C. Kan, M. Lenzner, P. Wobrauschek, and F. Krausz, *Science* **278**, 661 (1997).

[80] Z. Chang, A. Rundquist, H. Wang, M. M. Murnane, and H. C. Kapteyn, *Phys. Rev. Lett.* **79**, 2967 (1997).

[81] J. Seres, E. Seres, A. J. Verhoef, G. Tempea, C. Strelill, P. Wobrauschek, V. Yakovlev, A. Scrinzi, C. Spielmann, and F. Krausz, *Nature* **433**, 596 (2005).

[82] E. A. Gibson, A. Paul, N. Wagner, R. Tobey, D. Gaudiosi, S. Backus, I. P. Christov, A. Aquila, E. M. Gullikson, D. T. Attwood, M. M. Murnane, and H. C. Kapteyn, *Science* **302**, 95 (2003).

[83] P. B. Corkum, *Phys. Rev. Lett.* **71**, 1994 (1993).

[84] J. L. Krause, K. J. Schafer, and K. C. Kulander, *Phys. Rev. Lett.* **68**, 3535 (1992).

[85] K. J. Schafer, B. Yang, L. F. DiMauro, and K. C. Kulander, *Phys. Rev. Lett.* **70**, 1599 (1993).

[86] P. Colosimo, G. Doumy, C. I. Blaga, J. Wheeler, C. Hauri, F. Catoire, J. Tate, R. Chirla, A. M. March, G. G. Paulus, H. G. Muller, P. Agostini, and L. F. DeMauro, *Nat. Phys.* **4**, 386 (2008).

[87] Y. Fu, H. Xiong, H. Xu, J. Yao, Y. Yu, B. Zeng, W. Chu, X. Liu, J. Chen, Y. Cheng, and Z. Xu, *Phys. Rev. A* **79**, 013802 (2009).

[88] H. Xiong, H. Xu, Y. Fu, J. Yao, B. Zeng, W. Chu, Y. Cheng, Z. Xu, E. J. Takahashi, K. Midorikawa, X. Liu, and J. Chen, *Opt. Lett.* **34**, 1747 (2009).

[89] W. Schwanda, K. Eidmann, and M. C. Richardson, *J. X-Ray Sci. Technol.* **4**, 8 (1993).

[90] M. B. Gaarde, J. L. Tate, and K. J. Schafer, *J. Phys. B* **41**, 132001 (2008).

[91] C. Vozzi, F. Calegari, F. Frassetto, L. Poletto, G. Sansone, P. Villoresi, M. Nisoli, S. De Silvestri, and S. Stagira, *Phys. Rev. A* **79**, 033842 (2009).

[92] C. Jin, G. Wang, A.-T. Le and C. D. Lin, *Scientific Reports* **4**, 7067 (2014).

[93] S.-W. Huang, G. Cirmi, J. Moses, K.-H. Hong, S. Bhardwaj, J. R. Birge, L.-J. Chen, E. Li, B. J. Eggleton, G. Cerullo, F. X. Kärtner, *Nature Photon.* **5**, 475 (2011).

[94] A. Wirth, M. T. Hassan, I. Grguraš, J. Gagnon, A. Moulet, T. T. Luu, S. Pabst, R. Santra, Z. A. Alahmed, A. M. Azzeer, V. S. Yakovlev, V. Pervak, F. Krausz, and E. Goulielmakis, *Science* **334**, 195 (2011).

[95] S. Haessler, T. Balčiunas, G. Fan, G. Andriukaitis, A. Pugžlys, A. Baltuška, T. Witting, R. Squibb, A. Zaïr, J. W. G. Tisch, J. P. Marangos, and L. E. Chipperfield, *Phys. Rev. X* **4**, 021028 (2014).

[96] M.-C. Chen, C. Mancuso, C. Hernández-García, F. Dollar, B. Galloway, D. Popmintchev, P.-C. Huang, B. Walker, L. Plaja, A. A. Jaroń-Becker, A. Becker, M. M. Murnane, H. C. Kapteyn, and T. Popmintchev, *PNAS*, E2361 (2014).

[97]   B. Shan, S. Ghimire, and Z. Chang, *J. Mod. Opt.* **52**, 277 (2005).

[98]   M. Nisoli, S. D. Silvestri, and O. Svelto, *Appl. Phys. Lett.* **68**, 2793 (1996).

[99]   M. Nisoli, S. D. Silvestri, O. Svelto, R. Szipöcs, K. Ferencz, C. Spielmann, S. Sartania, and F. Krausz, *Opt. Lett.* **22**, 522 (1997).

[100] K. S. Budil, P. Salières, A. L'Huillier, T. Ditmire, and M. D. Perry, *Phys. Rev. A* **48**, R3437 (1993).

[101] O. Tcherbakoff, E. Mével, D. Descamps, J. Plumridge, and E. Constant, *Phys. Rev. A* **68**, 043804 (2003).

[102] A. Zaïr, O. Tcherbakoff, E. Mével, E. Constant, R. López-Martens, J. Mauritsson, P. Johnsson, and A. L'Huillier, *Appl. Phys. B* **78**, 869 (2004).

[103] I. Sola, E. Mével, L. Elouga, E. Constant, V. Strelkov, L. Poletto, P. Villoresi, E. Benedetti, J.-P. Caumes, S. Stagira, C. Vozzi, G. Sansonem and M. Nisoli, *Nat. Phys.* **2**, 319 (2006).

[104] K. Kondo, Y. Kobayashi, A. Sagisaka, Y. Nabekawa, and S. Watanabe, *J. Opt. Soc. Am. B* **13**, 424 (1996).

[105] N. Dudovich, O. Smirnova, J. Levesque, Y. Mairesse, M. Y. Ivanov, D. M. Villeneuve, and P. B. Corkum, *Nat. Phys.* **2**, 781 (2006).

[106] I. J. Kim, C. M. Kim, H. T. Kim, G. H. Lee, Y. S. Lee, J. Y. Park, D. J. Cho, and C. H. Nam, *Phys. Rev. Lett.* **94**, 243901 (2005).

[107] J. Mauritsson, P. Johnsson, E. Gustafsson, A. L'Huillier, K. J. Schafer, and M. B. Gaarde, *Phys. Rev. Lett.* **97**, 013001 (2006).

[108] H. Merdji, T. Auguste, W. Boutu, J.-P. Caumes, B. Carré, T. Pfeifer, A. Jullien, D. M. Neumark, and S. R. Leone, *Opt. Lett.* **32**, 3134 (2007).

[109] L. E. Chipperfield, J. S. Robinson, J. W. G. Tisch, and J. P. Marangos, *Phys. Rev. Lett.* **102**, 063003 (2009).

[110] H.-C. Bandulet, D. Comtois, E. Bisson, A. Fleischer, H. Pépin, J.-C. Kieffer, P. B. Corkum, and D. M. Villeneuve, *Phys. Rev. A* **81**, 013803 (2010).

[111] A. Fleischer and N. Moiseyev, *Phys. Rev. A* **74**, 053806 (2006).

[112] M. B. Gaarde, P. Antoine, A. Persson, B. Carré, A. L'Huillier, and C.-G. Wahlstrom, *J. Phys. B* **29**, L163 (1996).

[113] M. B. Gaarde, A. L'Huillier, and M. Lewenstein, *Phys. Rev. A* **54**, 4236 (1996).

[114] E. Goulielmakis, V. Yakovlev, A. L. Cavalieri, M. Uiberacker, V. Pervak, A. Apolonski, R. Kienberger, U. Kleineberg, and F. Krausz, *Science* **317**, 769 (2007).

Chapter 2

Principles of HHG in Laser-produced Plasmas using MIR Pulses

2.1. Introduction

Generation of harmonics resulting from intense laser-atom interactions is of particular interest because of its fundamental importance and practical applications. Most harmonic generation studies have been performed using pump lasers in the near-visible or ultraviolet. This has meant that multiphoton processes could be observed only in tightly bound systems such as the rare gases, as was shown in the previous chapter. HHG in rare gases provides coherent, short pulse, tabletop XUV sources, with spectra extending into the water window. These sources have found applications in atomic and condensed matter physics [1]. They may also provide a route toward the generation of very short duration XUV pulses [2, 3]. However, experimental progress has been limited, at least in part, by the lack of suitable materials for measuring XUV pulse lengths. For example, standard second harmonic intensity autocorrelation requires frequency doubling materials which are not available at wavelengths below 400 nm. To more readily understand the physics of the HHG process, especially the temporal properties, it may be possible to use intense MIR light to generate HHG in the UV/visible spectrum, where they can be characterized using conventional techniques.

Strong-field multiphoton processes, such as harmonic generation, using long wavelength pump lasers and laser-produced plasmas have until 1999 been virtually unexplored. The first experimental and theoretical investigation of HHG by MIR excitation in alkali metal vapors, specifically potassium and rubidium, was reported in [4]. In that work, harmonics extending to the 19th order of the fundamental field were observed.

In isotropic media, HHG using moderate-level femtosecond laser pulses allows easy production of coherent radiation in the XUV spectral range. During the last 25 years, predominantly rare gases were employed as target media for HHG, which, however, imposed some physical and practical limits on the performance of these coherent XUV sources. So far, only low conversion efficiencies of HHG have been reported using gases as the nonlinear media, despite enormous efforts. For practical applications of high-order harmonic sources, higher conversion efficiency and thus an increase in the photon flux and also of the maximum photon energy of the harmonic radiation would be beneficial. The generation of high-order harmonics in another isotropic medium, laser-produced plasma, being for this purpose a relatively new and largely unexplored medium, promises to yield these expectations, as well as to open the door for new developments in laser–matter interactions.

The application of laser plasma came up at the beginning of the nineties with the aim of optimizing HHG light sources. The idea was based on the use of the larger ionization potentials of alkali ions as compared to noble gases. The use of such plasma was aimed at improving the HHG phase matching conditions, and increasing the concentration of excited ions and neutrals in plasma, which was thought to be the right way for amendments of harmonic yield. In the meantime, the first experiments on HHG in the passage of laser radiation through a plasma produced during laser ablation of a solid target turned out to be much less successful. Data obtained with the use of highly excited plasmas containing multiply charged ions revealed several limiting factors, which did not permit generation of harmonics of sufficiently high orders [5, 6]. Those investigations stopped after the demonstration of relatively low-order harmonics. These disadvantages, as well as low conversion efficiency (10^{-7}–10^{-6}), led to the erosion of interest in this HHG technique, especially in comparison with the achievements involving gas HHG sources.

The advantages of plasma HHG could largely be realized with the use of low-excited and weakly ionized plasma, because the limiting processes governing the dynamics of laser frequency conversion would play a minor role in this case. Attention was drawn to this feature early in the study of third-harmonic generation in weakly ionized plasma [7]. This assumption allowed the formulation of several recommendations for further advancement toward the development of efficient shorter wavelength sources based on the frequency conversion of laser radiation in various plasmas. A new history of plasma HHG studies based on this approach started in 2005 [8]. It was immediately followed by observation of extended harmonics and considerably higher conversion efficiencies ($>10^{-5}$), which became comparable with those reported in gas HHG studies. Over the following years, enormous improvements in the characteristics of this process were reported.

A substantial increase in the highest order of the generated harmonics, emergence of a plateau in the energy distribution of highest-order harmonics, high efficiencies obtained with several plasma formations, realization of resonance-induced enhancement of individual harmonics, efficient harmonic enhancement for plasma plumes containing clusters of different materials, and other properties have demonstrated the advantages of using specially prepared plasmas for HHG.

Recent developments of this technique have revealed many new approaches and achievements in harmonic generation from plasmas. Among them are the studies of harmonics from clusters using the ablation of commercially available nanoparticles, fullerenes, carbon nanotubes, resonance-enhanced features of odd and even harmonics, generation of extremely broadband high-order harmonics, application of high pulse repetition rate laser sources and ultrashort pulses for HHG in plasma plumes, observation of quantum path signatures in the harmonic spectra from various plasmas, enhancement of harmonics from in-situ produced clusters, development of theoretical approaches describing the observed peculiarities of resonance-enhanced harmonics, emergence of a "second" plateau in harmonic distribution, development of two-color pump schemes for plasma-induced harmonics, observation of extremely strong HHG in carbon-contained plasmas, proposals for quasi-phase matching in plasma plumes, observation of the attosecond nature of pulse duration of plasma-induced harmonics, etc. [9–11].

In this chapter, we introduce the technique of the HHG in laser-produced plasmas using MIR driving pulses and show the advantages in application of the long-wavelength tunable sources of coherent radiation for the HHG using various approaches. In particular, we concentrate on the application of the high pulse repetition rate (1 kHz) MIR source for these purposes. We also compare the use of MIR and conventional (800 nm) sources of harmonic generation.

2.2. High-Order Harmonic Generation in Graphite Plasma Plumes Using 800 nm and MIR Laser Pulses

High-order harmonic generation in laser-produced plasmas is a promising method to produce coherent XUV radiation. The quest for ways to increase the HHG efficiency in plasma plumes in the XUV range remains a hot topic of strong-field laser-matter interaction studies. It has become obvious in recent years that the important requirement for efficient HHG is a specific composition of the laser plasma, which should contain predominantly neutral atoms and singly-charged ions. HHG studies that use highly excited plasmas containing multiply charged ions revealed several factors that limit the harmonic conversion efficiency and showed only a limited number of harmonic orders [12, 13].

Early studies show that several conditions have to be fulfilled to optimize the conversion efficiency during the interaction of the laser radiation with the plasmas [14]. Since the temporal evolution of a laser ablation plasma spans from several tens to hundreds of nanoseconds, it is necessary to synchronize the arrival of the laser pulse generating the harmonics (the "probe" pulse) with the moment when the plasma is 'optimal'. The term 'optimal' refers to the degree of ionization and excitation of atoms and ions, as well as the concentration of the laser plasma, which ensure efficient high-order nonlinear frequency up-conversion of the probe pulse propagating through the plasma. The position of the focus of the probe radiation with respect to different zones of the plasma is also of considerable importance. The optimization of frequency up-conversion is accomplished by adjusting the distance between the target surface and the propagating probe beam and by optimizing the delay between the ablating and probe pulses.

The characteristics of the laser plasma play a crucial role in determining how efficiently high harmonics can be generated in the plasma plumes. An increase in the free electron density was likely to have been the limiting factor for the harmonic cut-off energy in early experiments with laser plasmas. A search for appropriate target materials which can provide favorable ablation plasmas for efficient HHG has motivated the analysis of plasma characteristics at conditions of high harmonic yield. Recent studies have shown that carbon ablation plasmas are promising media to satisfy the above requirements [15–18].

Shot-to-shot stability of the harmonic signal is crucial for any application of the generated radiation and also for the measurement of the pulse duration of converted XUV radiation. Such temporal measurements were reported in the case of HHG in a chromium plasma [19]. It was shown that the 11th to the 19th harmonics of a Ti:sapphire laser form, in the time domain, an attosecond pulse train. It was underlined that instability of the harmonic signal in their experiments using a 10 Hz pulse repetition rate laser was the main obstacle for an accurate measurement of the temporal structure of plasma harmonics. Beside its fundamental interest, high-order harmonic generation in plasma plumes could thus provide an intense source of femtosecond and attosecond pulses for various applications.

Optical parametric amplifiers operating in the MIR range are promising tools for harmonic cut-off extension and attoscience experiments. The spectral cut-off of HHG obeys the scaling law $H_c \sim I\lambda^2$ [20], where I is the peak intensity of the probe field and

λ its central wavelength, which allows one to extend the harmonic emission beyond the 100 eV range by using longer wavelength laser sources. Another advantage of mid-infrared optical parametric amplifiers (MIR OPAs) is their wavelength tunability, which allows one to tune the spectral position of harmonics towards the ionic transitions with strong oscillator strengths. This feature allows the observation of resonance-enhanced harmonics and broadens the range of plasma samples where this phenomenon could be realized compared with the case of $\sim$800-nm lasers of essentially fixed wavelength. Moreover, by using two-color HHG techniques, the application of MIR OPAs allows the study of complex molecules during their ablation and HHG using the tunable long-wavelength radiation. These features are interesting for spectroscopic applications of HHG in the MIR range [21, 22].

The use of MIR OPAs for HHG also leads to a reduced harmonic generation efficiency that scales as λ^{-5} [23, 24]. It is of considerable interest to analyze the relative behavior of plasma harmonics in the cases of $\sim$800 nm and MIR lasers and thereby to find the conditions when the reduction of harmonic yield becomes not so dramatic due to some enhancement mechanisms, such as the presence of in-situ produced nanoparticles, which increase the HHG conversion efficiency. It is worth noting that previous studies of plasma HHG in carbon plumes have inferred, through analysis of plasma debris morphology, the formation of nanoparticles during laser ablation of graphite targets.

Atomic carbon is a reactive species which stabilizes in various multi-atomic structures with different molecular configurations (allotropes). All the allotropic forms of carbon (graphite, diamond, and amorphous carbon) are solids under normal conditions, but graphite has the highest thermodynamic stability. Laser ablation of graphite has been intensively examined during the last ten years to define plasma conditions for the synthesis of carbon structures with unique properties, in particular, fullerenes and carbon nanotubes. The physical characteristics of the plasma plume, such as concentration of atoms and clusters, directly affect the properties of the material being formed in the dynamic expansion of the ablated material. The successful synthesis of clusters is strongly dependent on the formation of atomic and molecular species with the required chemistry and aggregation ability. Thus, to select the optimal plasma conditions for HHG, a detailed understanding of the basic physical processes governing the ablation plume composition and reliable methods for controlling of the plume species are needed.

The reasons mentioned above and the consideration of previous studies of HHG in carbon plasmas, as well as reported comparisons of the HHG in graphite-ablated plasmas and argon gas [17,18], have prompted us to systematically analyze the plasma conditions for optimal HHG conversion efficiency in graphite plasmas [25]. Herein we show the systematic analysis of HHG in graphite-ablated plasmas using ultrashort (3.5 and 30 fs) driving laser pulses. We also show efficient HHG frequency up-conversion of 3.5 fs Ti:sapphire laser pulses in the range of 15–26 eV using optimally prepared plasma plumes and demonstrate the tuning of harmonic spectra at variable conditions of the second stage of compression of the driving laser. We also present results on HHG in carbon plasmas using MIR driving pulses, which have allowed the extension of the harmonic cut-off while maintaining a comparable conversion efficiency with regard to the 780 nm driving radiation [26].

2.2.1. *Experimental arrangements*

The Ti:sapphire laser provided pulses of 25 fs and energies of up to 0.8 mJ at a repetition rate of 1 kHz. These pulses were focused into a 1-m-long differentially pumped hollow core fiber (250 μm inner core diameter) filled with neon. The spectrally broadened pulses at the output of the fiber system were compressed by 10 bounces of double-angle technology chirped mirrors. A pair of fused silica wedges was used to fine tune the pulse compression. High-intensity few-cycle pulses (760 nm central wavelength, 0.2 mJ, 3.5 fs, pulse repetition rate 1 kHz) were typically obtained in this system. The compressed pulses were characterized with a spatially encoded arrangement for direct electric field reconstruction by spectral shearing interferometry. This radiation was used as the probe pulses for frequency up-conversion in the specially prepared carbon plasma.

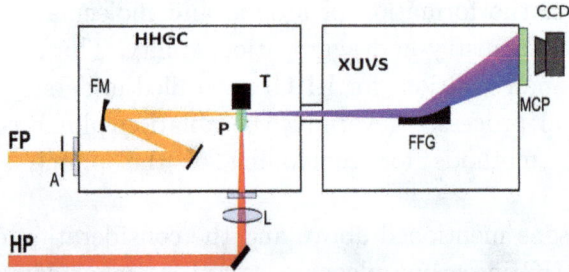

Fig. 2.1. Experimental setup for harmonic generation in plasma plumes. FP: femtosecond probe pulse, HP: picosecond heating pulse, A: aperture, HHGC: high-order harmonic generation chamber, FM: focusing mirror, L: focusing lens, T: target, P: plasma, XUVS: extreme ultraviolet spectrometer, FFG: flat field grating, MCP: microchannel plate and phosphor screen detector, CCD: CCD camera. Reproduced from [26] with permission from IOP Publishing.

A portion of the uncompressed radiation of this laser (central wavelength 780 nm, pulse energy 120 μJ, pulse duration 8 ps, pulse repetition rate 1 kHz) was split from the beam line, prior to the laser compressor stage, and was focused into the vacuum chamber to heat the graphite target and create a plasma on its surface (Fig. 2.1). The picosecond heating pulses were focused by a 400 mm focal length lens and created a plasma plume with a diameter of ∼0.5 mm using an intensity on the target surface of $I_{ps} = 2 \times 10^{10}$ W cm^{-2}. The delay between plasma initiation and femtosecond pulse propagation was fixed at 33 ns. As an alternative ablation laser, we also used 10 ns, 1064 nm pulses from a 10 Hz repetition rate Q-switched Nd:YAG, laser that provided an intensity on the target surface of 1×10^9 W cm^{-2}. In this case the delay between the 10 ns heating and the 3.5 fs probe pulses was varied in the range of 10 − 60 ns to maximize the harmonic yield.

The 3.5 fs probe pulses, propagating in a direction orthogonal to that of the heating pulse, were focused into the laser plasma using a 400 mm focal length reflective mirror. The position of the focus with respect to the plasma area was chosen to maximize the harmonic signal, and the intensity at the plasma area at these conditions was estimated to be $I_{fs} = 6 \times 10^{14}$ W cm^{-2}. We also used 30 fs, 780 nm, 2 mJ probe pulses from another Ti:sapphire laser operating at 1 kHz repetition rate and producing approximately the same intensity inside the laser plasma ($4 \times 10^{14}\,\mathrm{W\,cm^{-2}}$). The generated harmonics were analyzed by an XUV spectrometer consisting of a flat-field grating and a microchannel plate coupled to a phosphor screen. Images of harmonics were recorded by a CCD camera.

In order to analyze the harmonic yield of the MIR source in the graphite-ablated plasma an OPA pumped by the 30 fs Ti:sapphire laser was used. A beam splitter inserted before the laser compressor of this Ti:sapphire laser allowed us to pick off 10% of the beam (780 nm, 1 mJ, 20 ps, 1 kHz pulses) to generate a plasma plume on the graphite targets, with the remaining 90% being compressed to 30 fs (7 mJ) to pump a commercial computer-controlled OPA (HE-TOPAS, Light Conversion). The OPA was optimized for high conversion efficiency and short duration of the converted pulses. To achieve high reproducibility of the generated pulses, the last amplification stage was driven to saturation. This device generated signal pulses in the 1200–1550 nm range with a maximum power of 1.8 mJ at ∼1300 nm. The idler pulse covers the 1700–2200 nm range with a maximum power of 1 mJ at ∼2000 nm. Pulse duration has been characterized by a SPIDER type setup with about 40 fs pulse duration in the overall range. The delay between the heating ablation pulse and MIR pulses from the OPA was set to 35ns, as this delay was found to be optimal for the efficient generation of extended harmonics.

Plasma characterization through optical spectroscopic measurements in the visible, ultraviolet (UV) and XUV ranges at the conditions of different HHG efficiencies were carried out using the above-described XUV spectrometer and a fiber spectrometer. The fluences of heating pulses at which the spectra were recorded corresponded both to optimal and non-optimal conditions of HHG. The acquisition times were set to 1 s, for measurements of XUV spectra, and to 0.5 s, for measurements of visible and UV spectra. Characterization of the plasma debris collected on silicon wafers placed 4 cm from the ablated target was carried out by scanning electron microscopy (SEM). The details of experimental setup are available in [17].

2.2.2. *HHG at the conditions of stabilized plasma formation*

In this subsection, we analyze the studies of HHG from graphite-ablated plasma by 10 ns pulses operating at 10 Hz repetition rate. The shortest available probe pulses were used to analyze the various features of this process depending on the spatio-temporal parameters of the converting radiation. Harmonics from the carbon plasma were optimized by choosing the delay between pulses, and the distance between the femtosecond beam and the target. Previous analysis of low-order harmonic spectra using few-cycle pulses propagating at different delays with respect to a 10 ns pulse-induced plasma [17] have shown that delays in the 30–40 ns range yield the highest conversion efficiency and so we used delays in this range as a basis for further improvement of high harmonic yield. To analyze the influence of the spectro-temporal characteristics of the probe radiation on the harmonic yield, we changed the backing pressure of neon in the hollow fiber, which allowed the variation of pulse duration from 25 to 3.5 fs. The dependence of the spectral and intensity characteristics of the harmonic images recorded by the CCD camera in the 15–25 eV range at different input pulse spectra and backing pressures of neon are shown in Fig. 2.2. One can clearly see that with the increase of backing pressure (from 1.2 to 3 bar), the harmonic intensity increases, while the harmonic wavelength spectrally shifts towards the blue. During these experiments the driving pulse energy was held constant.

The measurements of the pulse durations were carried out for extreme cases, i.e. in the absence of neon gas in the fiber (0 bar, 25 fs) and in the filled fiber (3 bar, 3.5 fs). One can note that the pulse duration and the corresponding spectra did not change significantly at lower pressures of Ne. The strong SPM and spectral broadening appeared at pressures above 1.2 bar. Therefore it is safe to expect some scaling of pulse duration for the spectra presented on the left side of Fig. 2.2, which correspond to the scaling of spectral broadening.

An interesting feature of the carbon harmonic spectrum from the 10-ns-induced plasma is that the spectral width is about 2–3 times

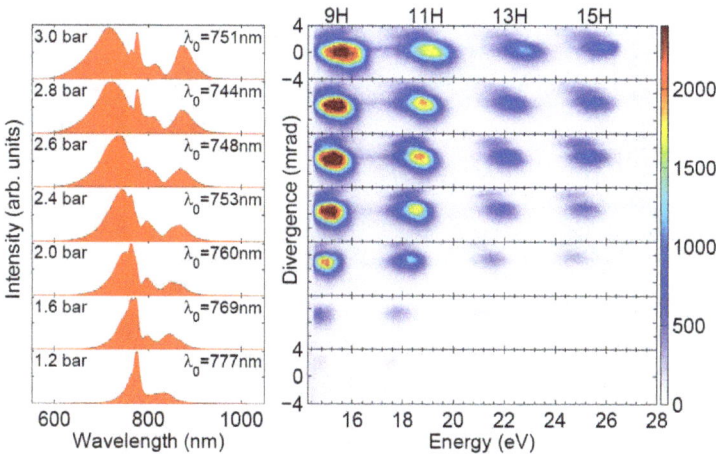

Fig. 2.2. Carbon harmonic spectra as a function of neon pressure in the hollow fiber. The corresponding probe pulse spectra measured in front of the vacuum chamber are presented on the left side. The plasma was created using the 10 ns pulses. λ_0 is the central weighted wavelength of the spectral distribution. The color scale indicates the harmonic intensity. Reproduced from [26] with permission from IOP Publishing.

broader than that of harmonics generated in other atom- and ion-rich plasmas at the same fluence and intensity of heating pulse, when using few-cycle pulses. For example, the full width at half maximum for medium-order harmonics was 1.5 nm in the case of graphite plasma, versus 0.4 nm for different metal (Ag, Al, and Cu) plasmas. The broader width of the harmonics can be explained by self-phase modulation and chirping of the fundamental radiation propagating through the carbon plasma. The presence of nanoparticles in the plasma plume may also contribute to bandwidth broadening of harmonics.

For practical applications of the coherent short-wave radiation generated in a graphite plasma using a 1 kHz driving laser, it is necessary to analyze the stability of the plasma characteristics and of the generated harmonics. In the experiments described below we used the 8 ps heating pulses operating at 1 kHz in an attempt to improve the stability of plasma harmonics at these conditions, which have previously been unfavorable for stable plasma formation. A new

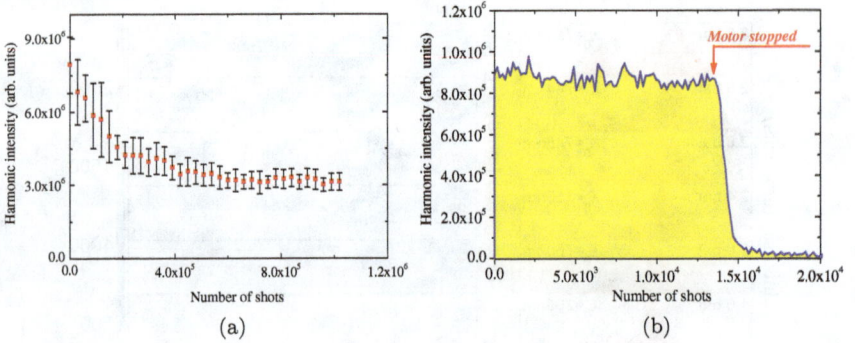

Fig. 2.3. Harmonic generation in graphite plasma using 3.5 fs pulses. (a) Stability of harmonic intensity over 1 million shots on a graphite target integrated over the 11th to 25th harmonics. (b) Decay of harmonics after stopping the motor rotating the target. Reproduced from [26] with permission from IOP Publishing.

technique for maintaining a stable ablation plasma for harmonic generation using high pulse repetition rate lasers (>1 kHz) based on ablation of rotating metal targets has been introduced in [27]. The discussed studies show that, in spite of the different properties of metal and graphite targets, the rotating target allowed stable HHG in both metal and graphite plasmas. Figs. 2.3a and 2.3b show the improved stability over $\sim 10^6$ laser shots of the 11th to 25th harmonics when using a rotating graphite target and how the harmonic intensity rapidly decays after the target rotation is stopped. The rotating graphite rod allows maintenance of a relatively stable harmonic yield well above 1×10^6 laser shots. Harmonics up to the 29th order were routinely observed in these studies using the 3.5 fs pulses.

It is worth noting that harmonic intensity is the same when returning to the same spot after one rotation of the graphite rod. This reveals the unchanged morphological target conditions. Indeed, target analysis by optical inspection has confirmed that there is negligible surface modification due to laser ablation provided we continuously move the ablation spot along the target surface. This means that in graphite, under repetitive ablation on the same target position at 1 kHz, the instability of the plasma and generated harmonics is largely related with the unstable conditions of the ablating spot. After moving to a new spot, the previous irradiated target area cools down and becomes again available (one rotation later) for further ablation with the same harmonic output. Therefore using a rotating graphite rod can significantly improve the stability of the harmonic signal.

The reason for the slow degradation of harmonic (Fig. 2.3a) is clearly due to effects on the surface of the target. Note that stable harmonics can be generated over a broad range of target rotation speeds. Given the target rotation speed as well as the size of the ablation focus, the same target area was repeatedly exposed to ablation for consecutive rotations over the 20 minute duration of our experiments. At the high pulse repetition rate this could lead to thermal damage of the target. It is possible that, once the solid surface is melted, the force from a following laser shot and from the subsequently created plasma causes the expulsion of part of the liquid target from the ablation area, creating a deeper hole out of the focus or with angled sides. These effects would result in plasma emission in a range of directions around the normal to the surface. They can be considerably diminished once the target starts to rotate. During rotation, the previously ablated area cools down such that, during the forthcoming ablation event of the same spot, the plasma formation occurs at approximately the same conditions.

To prove that the ablated area cools down with rotation, we rotated the target at different speeds (from 10 to 300 rpm) and did not find a difference in the stability of the harmonic yield. These observations point out the importance of the periodic renewal of ablation zone and confirm our suggestion that the cooling of the ablation area leads to stable plasma generation.

From the above consideration, one can conclude that the decay is mostly related with the creation of a micro-channel on the target surface. The separate studies of this phenomenon have shown that, in the case of metal targets, this decay becomes less pronounced compared with graphite. This difference is related to the thermodynamic properties of the latter material surface which cannot withstand a long series of ablation events without changing the surface properties. The details of these studies are presented in [27].

The averaging of harmonic intensity was carried out during each 0.5 minute (i.e. for 30,000 laser shots). One should note that, due to integration over the 15–26 eV spectral range of the intensity of a few harmonics, the fluctuations of this parameter could be attributed to variations of the free electrons density, which change the phase-matching conditions for the harmonics. The error bars reflect these variations over each 30 seconds of the measurement.

Here we analyze the results of HHG in carbon plasmas using the MIR OPA pulses (wavelength 1300 nm, pulse duration 40 fs, 1 kHz pulse repetition rate) as probe radiation and compare these results with those obtained with 780 nm, 30 fs driving pulses operating at the same pulse repetition rate. The upper panel in Fig. 2.4 shows the harmonic spectrum generated in the case of 1300 nm probe pulses. Harmonics up to the 57th order were observed at the conditions of carbon plasma formation using the uncompressed 20 ps pulses from this laser. It is worth noting that application of less intense 1400 nm pulses available by tuning the OPA, while generating weaker harmonics, did not result in a higher harmonic cut-off than in the case of 1300 nm. This observation suggests that the harmonic generation occurred under saturated conditions, with the expectation of even stronger harmonics once the micro- and macro-processes governing frequency conversion are optimized.

Harmonic spectra up to the 29th order in the case of 780 nm, 30 fs probe pulses, are presented in the bottom panel of Fig. 2.4. By comparing with the spectra collected with the 1300 nm driving source (same figure, upper panel), one can clearly see the expected extension of harmonic cut-off in the case of the longer-wavelength driving source. The important peculiarities of these comparative studies are the broadband harmonic spectra in the case of 1300 nm laser and the similar yield of harmonics at the two driving wavelengths. Whilst the former feature depends on the bandwidth of the OPA output, the latter observation requires additional consideration. It was observed that the plasma harmonic yield from the MIR source did not follow the expected $I_h \propto \lambda^{-5}$ rule. In fact, for the intensities of MIR and 780 nm pulses used [$\sim (2-4) \times 10^{14}$ W cm^{-2}], the harmonic efficiency of the XUV radiation driven by MIR pulses was higher

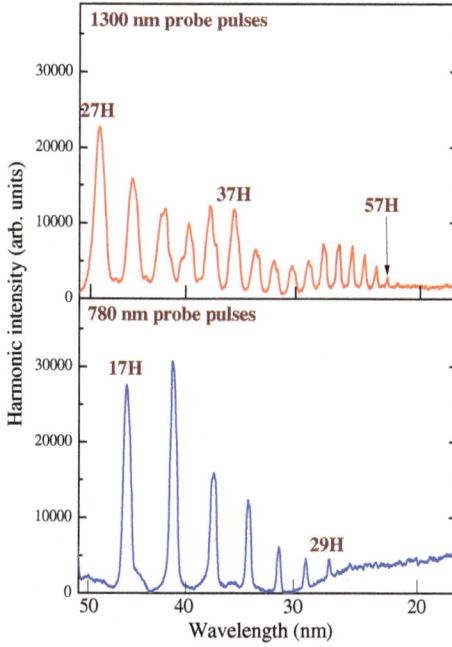

Fig. 2.4. Plasma harmonic spectra using the 1300 nm (upper panel) and 780 nm (bottom panel) probe pulses. The energies of probe pulses were 0.2 mJ (upper panel) and 0.54 mJ (bottom panel). Ablation was carried out using 20 ps, 780 nm, 1 kHz laser pulses. Reproduced from [26] with permission from IOP Publishing.

compared with the case of 780 nm pulses, assuming lower energy of the former pulses (0.2 and 0.54 mJ respectively). One can note that the $I_h \propto \lambda^{-5}$ rule predicts a $\sim$13-fold decrease of conversion efficiency for the MIR (1300 nm) pulses compared with the 780 nm pulses at equal probe intensity.

2.2.3. *Discussion*

A few earlier studies have suggested that the presence of nanoparticles in carbon laser ablation plasmas can explain the observed strong harmonic yield from these media [15, 16]. It was reported that the debris from ablated graphite and carbon targets contained nanoparticles with sizes between 100 and 300 nm. The authors of these studies therefore suspected that nanoparticles formed in the plasma by ablation were the source of intense harmonics. Heterogeneous decomposition, liquid phase ejection and fragmentation, homogeneous nucleation and decomposition, and photomechanical ejection are among the processes that can lead to the production and disintegration of nanoparticles [28–31]. A number of different techniques were used in these studies to determine the aggregation state of the evaporated material, including time-resolved emission spectroscopy, CCD camera imaging of the plasma plume, Rayleigh scattering, and laser-induced fluorescence.

In the discussed studies, SEM for debris analysis and time-of-flight mass spectrometry (TOFMS) of plasma characterization were used. These two methods have provided useful clues about the conditions and dynamics of plasma plume formed above the target surface. Whilst the former method can provide information about the presence of nanoparticles in the plasma, one has to cautiously consider those results from the following point of view. The deposition process on the substrate happens much later than the time of HHG emission, and the physical process of deposition may lead to further aggregation. Since SEM is an ex-situ method, one cannot exclude the difference between the real composition of clusters in the plasma and the results of SEM measurements, although it clearly proves the presence of clusters in the plasma. TOFMS yields information on the *in-situ* presence of ionized clusters, although it requires ablation of the target at the same conditions as in the case of HHG experiments and is not well suited for the detection of neutral nanoparticles in the ablated plasma.

The TOFMS measurements did not reveal the presence of neutral clusters in the 10 ns pulse produced plasma for the reasons described in the previous section. However, other studies (see for example [32]) have indicated the presence of neutral carbon clusters using two- and probably one-photon ionization with an ArF laser (photon energy 6.4 eV). Early TOFMS studies of laser ablation of graphite have revealed the typical characteristics of the expanding plasma species (average velocity 1.5×10^5 cm s^{-1}) and their concentration ($4 \times 10^{18} - 6 \times 10^{19}$ cm^{-3} [33]) for ablation with 532 nm, 10 ns pulses at fluences of the order of 3 J cm^{-2}. These values are comparable to the results of the calculation presented in 3.2.4 of plasma concentration using picosecond and nanosecond heating pulses ($2 \times 10^{17} - 2 \times 10^{18}$ cm^{-3}), where the presence of clusters in the plasma plume has been demonstrated.

The measured mass distributions revealing the presence of C_{10} to C_{30} species, is in good agreement with those observed in previous studies of graphite-ablated plasma at similar excitation conditions [33]. The restriction of cluster sizes to small-sized carbon nanoparticles has also been reported in [34], where it was argued that stronger excitation conditions are necessary to observe clusters larger than C_{32}. In that case, one should expect the appearance of closed cages made of joined five and six member rings. It was confirmed that C_{60} and C_{70} fullerenes are the most abundant species among the high-mass ions of the carbon plasma plumes at higher ablation fluences. It was also suggested [33] that it is very likely that the plasma is sufficiently dense for cluster growth to occur via ion-molecule reactions. The kinetic mechanism can be responsible for the formation of carbon cluster ions since the supersonic entrainment method is expected to considerably cool down the cluster ions. The growth of clusters is based on the addition of many small carbon neutral species to the ions in a stepwise fashion. One must bear in mind that while these assumptions are correct for specific conditions of over-excited graphite targets, during our TOFMS measurements we maintained conditions of "soft" laser ablation to ensure efficient HHG.

An explanation for strong harmonic generation from nanoparticles compared with single atoms or ions could be the higher concentration of neutral atoms inevitably accompanying the presence of nanoparticles. Unlike single atoms and ions, whose density quickly decreases due to plasma expansion, nanoparticles retain local densities that are close to solid state. The increase of recombination cross-section for clusters with respect to atoms can also potentially enhance the HHG efficiency in nanoparticle-contained plasmas. Earlier studies of HHG from gases [35, 36], as well as from plasmas containing various nanoparticles (Ag, Au, $BaTiO_3$, etc) [37], have proven these assumptions by demonstrating the enhanced HHG from clusters as compared with single atoms and ions. Further evidence of the cluster contribution to the enhancement of the harmonic generation process comes from investigations of very intense laser ablation of a silver target [38] which gave clues regarding the participation of in situ generated nanoparticles.

The observation of a strong extended harmonic plateau in the case of the 1300 nm probe radiation also suggests the involvement of clusters in the HHG process with MIR pulses. Assuming the expected decrease of harmonic intensity from single particle emitters with the growth of driving radiation wavelength ($I_h \propto \lambda^{-5}$, [39]), one can anticipate at least one order of magnitude decrease of harmonic yield from MIR pulses as compared with the yield obtained with 780 nm radiation at other equal conditions, in particular, pulse energy and duration. However, the experiment did not show a considerable difference between the intensities of harmonics that originated from these two driving sources (Fig. 2.4). The energy of the 1300 nm pulses in the plasma area (0.2 mJ) was lower than the Ti:sapphire pulse (0.54 mJ). This suggests the involvement of a mechanism which compensates for the expected considerable decrease of harmonic efficiency for the longer-wavelength laser. The involvement of a clustered component of the laser plasma in the process of frequency up-conversion could arguably explain the observed inconsistency with the theoretical predictions of the $I_h \propto \lambda^{-5}$ rule.

In principle, the intensity enhancement of the harmonic spectrum from the carbon plume in the 15–26 eV range implies the involvement of surface plasmon resonances of nanoparticles, analogously to the case of fullerenes [40] in the range of their giant resonance in the vicinity of 20 eV. To prove this in the case of carbon plasma, one should provide evidence of giant absorption in the above range, but this has not been reported yet in the literature. The plasmonic properties of carbon nanoparticles can be responsible for the observed enhancement of carbon harmonics, however their role requires additional study. Another option for explaining the high harmonic generation yield in the carbon plume is the indirect involvement of the clusters in HHG, that, while not participating as harmonic emitters, could rather enhance the local field, analogously to recently reported studies using gold nanostructures enhancing gas HHG [41].

The discussed experimental studies revealed the limitations of target excitation for these two regimes of plasma formation (8 ps and 10 ns). However, it is possible to apply the advantages of the ablation process by generating larger plasma plumes using a cylindrical optics and maintaining the same optimal intensities (2×10^{10} and $1 \times 10^9 \, \mathrm{W \, cm^{-2}}$ for 8 ps and 20 ns pulses) along the whole extended area of ablation. In that case one can expect further enhancement of harmonic yield caused by the increase of length of the nonlinear medium.

These studies have presented evidence of the superior properties of graphite ablation for HHG. Some arguments which could explain the enhanced high harmonic yield from this medium are as follows: (a) the graphite target allows easier generation of a relatively dense carbon plasma and the production of adequate phase-matching conditions for lower-order harmonic generation, (b) the first ionization potential of carbon is high enough to prevent the appearance of high concentration of free electrons, a condition that is not necessarily met in metal plasma plumes, (c) neutral carbon atoms dominate in the carbon plume at optimal conditions of HHG before the interaction with the femtosecond laser pulse, and (d) carbon species allow the formation of multi-particle clusters during laser ablation, which can enhance the HHG yield.

2.3. MIR- and 800-nm-Induced High-Order Harmonic Generation in Uracil Laser Plumes

2.3.1. *Introduction*

As it was already shown, a possible outcome of the interaction of atoms with intense, ultrashort laser pulses is the emission of coherent XUV photons [42]. During the interaction, the strong electric field of the laser pulse causes an atomic electron to tunnel and drive away from its parent ion. Once the field reverses, the electron is slowed down to a stop and then reaccelerated in the opposite direction towards the parent ion. There is a possibility that the electron will recombine with the host and return to the electronic state it originally tunnelled out of, resulting in the emission of the coherent short wavelength photons. The energy of these photons will be equal to the sum of the kinetic energy gained during reacceleration and the ionisation potential of the state it tunnelled out of.

Previous studies of laser plasma HHG have been mostly confined to ablation targets constituted by a single element, such as graphite or silver, or by two-elements targets such as gallium arsenide [43]. The extension of the technique to more complex species including biological molecules is of high interest, as these studies can unravel some electronic dynamics that take part in highly relevant, and not yet well understood processes of photosynthesis or DNA damage. To validate the application of laser plasma HHG spectroscopy to large molecules, the ribonucleic acid (RNA) base uracil has been examined and a comparison with its deoxyribonucleic acid (DNA) counterpart, thymine, has been established [44].

Uracil ($C_4H_4N_2O_2$) is a naturally occurring pyrimidine derivative and one of the four nucleobases in RNA where uracil binds to adenine via two hydrogen bonds. Methylation of uracil produces thymine ($C_5H_6N_2O_2$), which replaces uracil in DNA. The similar molecular structures of uracil and thymine are shown in Fig. 2.5. At room temperature uracil and thymine are both solids with melting points of 335°C and 317°C respectively. The low-lying vertical ionization potential of uracil is 9.5 eV while that of thymine is 9.1 eV [45].

In the refereed work, HHG from laser ablation plasmas of uracil and thymine targets using heating pulses of 1064 nm, 10 ns and of 780 nm, 160 ps lasers was examined. Harmonics were observed from uracil plasma using both 780 nm and 1300 nm driving radiation with pulses of 30 fs under the two laser ablation regimes. In contrast, the nonlinear optical response of thymine appeared to be quite different, as no signs of HHG were observed. Analysis of the ablation plumes of the two compounds in different spectral ranges and conditions of plasma formation, together with consideration of their chemical structure, allow discussion of the obtained results. These studies constitute the first attempt to analyse differences in structural

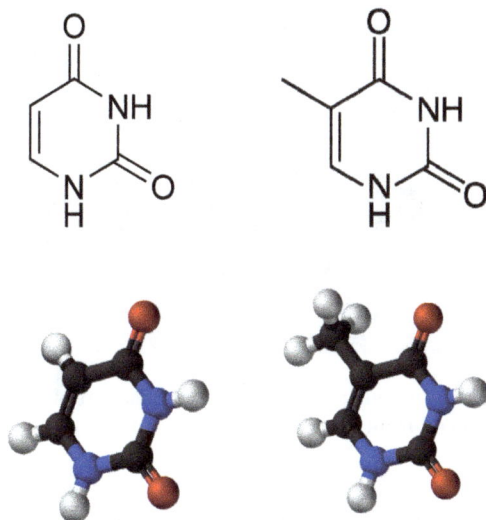

Fig. 2.5. Molecular and orientational structures of uracil (left column) and thymine (right column). Reproduced from [44] with permission from PCCP.

properties of complex molecules through plasma ablation-induced HHG spectroscopy.

2.3.2. *Experiment and discussion*

The experimental scheme was described in details in subsection 2.2.1. A part of the uncompressed radiation of Ti:sapphire laser was split off from the beam line prior to the compressor stage (780 nm, 1.3 mJ, 160 ps) and used for ablation of the solid targets placed in a vacuum chamber. After a delay of ∼60 ns, required for formation and expansion of the plasma plume away from the surface of the target, the compressed laser pulse (780 nm, 1 mJ, 30 fs) was focused onto the plasma using a 200 mm focal length mirror at an intensity of 4×10^{14} W cm^{-2} to generate the high-order harmonics. As an alternative ablation laser, the 10 ns, 1064 nm pulses from a 10 Hz repetition rate Q-switched Nd:YAG laser was also used. In that case the delay between the 10 ns heating pulse and the 30 fs probe pulse was varied in the range of 10–80 ns to maximize the harmonic yield. HHG

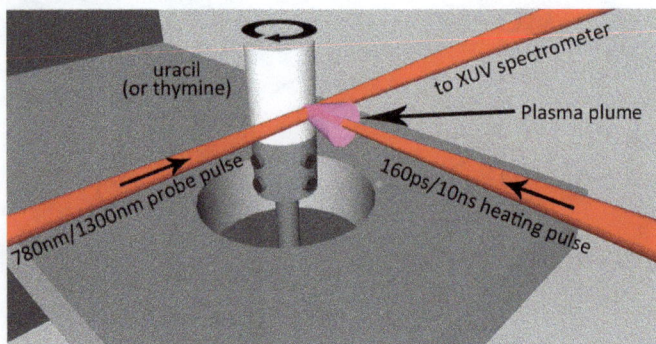

Fig. 2.6. Experimental setup of plasma plume ablation and high harmonic generation. Reproduced from [44] with permission from PCCP.

experiments were also performed using MIR radiation as a driving laser source (see subsection 2.2.1 for experimental description).

Uracil and thymine targets were prepared by stamping the powdered materials into solid pellets using a die and fly press. This method ensured a smooth surface to be exposed to laser ablation. For some experiments, targets containing 50:50 weight ratio of both compounds were also prepared. The targets were rotated (Fig. 2.6) to minimize overheating and damage of their surface from repeated laser shots, thus ensuring more stable ablation conditions [27]. Characterization of the plasma debris collected on silicon wafers, placed 4 cm from the ablated target, was carried out by SEM. The spectral studies of atomic and ionic emission from the laser-produced plasmas were carried out in the visible range (475–600 nm). The spectral characteristics of laser plasma in the visible and UV ranges were analyzed using the fiber spectrometer.

The problem of using the powder-containing targets is the shot-to-shot instability and rapid decrease of the harmonic yield in the case of static samples, due to the abrupt change in the target morphology following ablation. Such instabilities limit the applications of powdered materials for plasma HHG spectroscopy. The use of rotating targets containing powders could be considered as a method for improvement of the harmonic stability from the plasma produced on the powder-containing surfaces, which was implemented in this work. It was found that once the target rotation is stopped, the harmonic efficiency from the plasma decreased to the noise level within 1–2 seconds.

This target fabrication method could be very useful in the case of any powder-like species (fullerenes, metal nanoparticles, nanotubes, organic powder-like samples, etc). Previously, even at 10 Hz pulse repetition rate, the ablation of static powdered targets led to their rapid degradation and abrupt decrease of harmonic yield [46]. In the case of powder-like targets ablated by repetitive laser shots on the same spot at 1 kHz, the degradation of plasma plume is largely related with the unstable conditions of the ablating spot where the appearance of melted bath can considerably change the conditions of optimal plasma formation. After moving to a new spot, the previously irradiated target area cools down and becomes available again (one rotation later) for further ablation with approximately the same harmonic output. One can note that the stability of harmonics from such powder-like targets is still worse than from rotating bulk metal rods due to the faster degradation of the target surface in the former case.

Harmonics were generated from the uracil plasma plume using 30 fs driving laser pulses at 780 nm and 1300 nm radiation (Fig. 2.7). One can see that the harmonics generated at the two wavelengths are of comparable intensity, even if the energy in the 1300 nm pulse is smaller than that of the 780 nm pulses (0.35 mJ and 0.53 mJ respectively). This pattern of intensities does not follow the expected wavelength scaling rule of harmonic intensity $I_h \propto \lambda^{-5}$, which, for this case, would predict a ~13-fold decrease in conversion efficiency for the MIR source under equal experimental conditions. This could suggest the involvement in HHG of non-atomic entities, such as uracil fragments or clusters forming in the plasma plume, a scenario for which the above scaling rule would not apply.

Harmonic spectra from uracil were studied as a function of the wavelength of the MIR radiation. As shown in Fig. 2.8, this allowed tuning the emission from harmonics across a broad spectral range. In particular, the 35th harmonic could be tuned from 35 to 43 nm when the MIR driving source covered the 1200–1500 nm range.

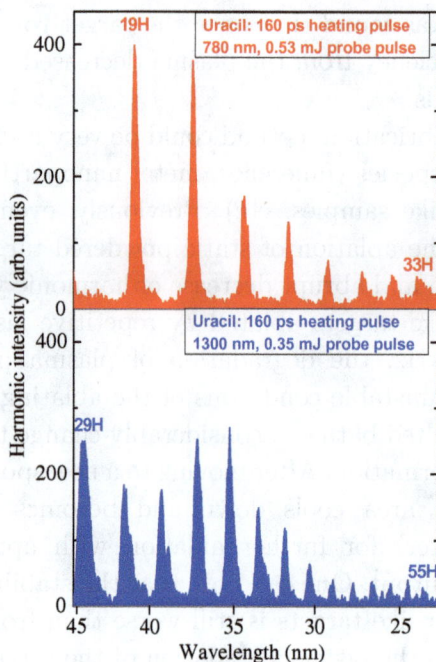

Fig. 2.7. HHG in uracil using 780 and 1300 nm probe pulses and 780 nm, 160 ps ablating pulses. Reproduced from [44] with permission from PCCP.

The harmonic cut-off in the case of MIR source was above the energy corresponding to the 65th order, while in the case of 780 nm radiation the harmonics up to the 39th order were generated.

Uracil was compared with its DNA equivalent thymine in various sets of experiments. The results in the case of 780 nm driving pulses are presented in Fig. 2.9(a). The pure uracil spectra (upper panel) clearly showed high-order harmonics while no such emission was detected from the thymine plume. Despite repeated attempts at different ablation intensities, using both the ns and ps pulses and delays, we were unable to observe any HHG signal from the thymine ablation plume. The 50:50 mixture of the two compounds (lower panel) displayed the same harmonic spectra as pure uracil, although with lower intensity, as expected, given that thymine plume species do not seem to participate in harmonic generation.

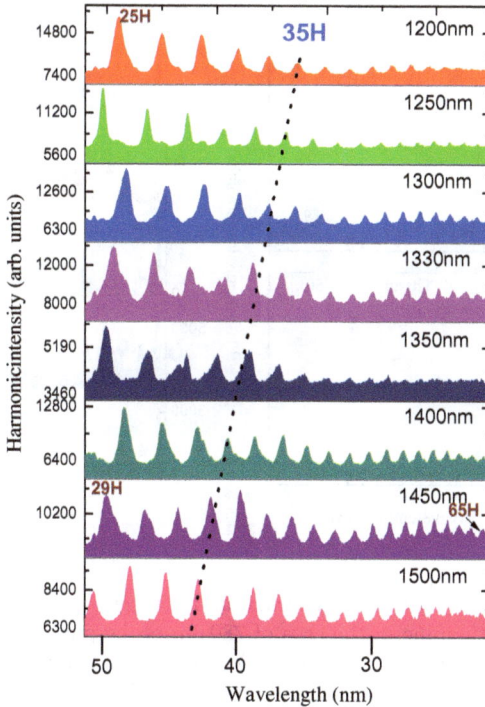

Fig. 2.8. HHG spectra from uracil obtained by tuning the central wavelength of the MIR driving pulse between 1200 nm and 1500 nm. The 35th harmonic order is tuned in the range of 35–43 nm. Ablation of the target was carried out with 1064 nm, 10 ns laser.

From the above it is clear that differences in composition between the ablation plasmas of the two pyrimidine nucleobases uracil and thymine are likely responsible for the different high order harmonic response. To understand this better, emission spectra obtained from the targets ablated at high pulse ablation intensities, leading to over-excitation of the targets, were collected. These are displayed in Fig. 2.9(b). Upon 1064 nm, 10 ns ablation, the spectra from uracil and thymine are rather similar. Ablation at 780 nm, 160 ps leads to emission spectra with a higher intensity in the short wavelength region, as compared with those obtained upon ns ablation.

Further analysis of the laser plasma plumes was carried out to measure the change in emission spectra with irradiation time,

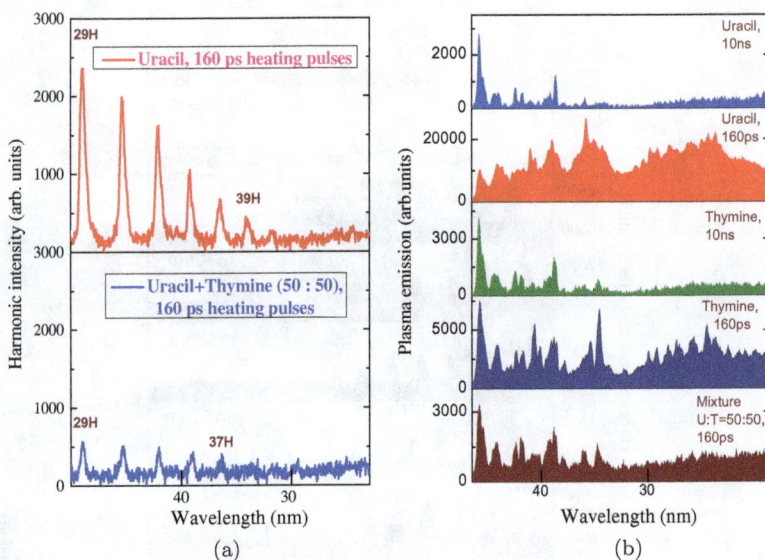

Fig. 2.9. (a) XUV spectra from uracil and a target made of a 50:50 mixture of the uracil and thymine in the case of 780 nm probe pulses and 780 nm, 160 ps ablating pulses ($I = 2 \times 10^{10}\,\mathrm{W\,cm^{-2}}$). Adapted from [44] with permission from PCCP. (b) XUV plasma emission from over-excited uracil, thymine, and a 50:50 mixture ablated using both 1064 nm, 10 ns and 780 nm, 160 ps ablating pulses. These spectra were measured at $5 \times 10^{10}\,\mathrm{W\,cm^{-2}}$ intensity of 160 ps ablating pulses and $3 \times 10^{9}\,\mathrm{W\,cm^{-2}}$ intensity of 10 ns ablating pulses.

when ablation was performed on the same spot of a static target. It was observed that the intensity of emission lines relative to a continuous background decreases upon repetitive laser exposure. The faster intensity decrease in thymine is indicative of the lesser stability of thymine targets as compared with uracil.

These studies are the first attempt to analyze the influence of structural properties of similar molecular compounds through plasma ablation-induced HHG spectroscopy. From the comparison of the two pyrimidine nucleobases uracil and thymine, which only differ by the methylation of one of the ring carbon atoms, some conclusions can be drawn about their specific structural and/or optical properties. While in the case of uracil efficient HHG was observed in the ablation plume of this compound, HHG from thymine wasn't observed in a wide range of explored experimental conditions. These studies have shown that the differences in composition of the respective ablation plumes are responsible for the strongly dissimilar nonlinear behavior during the interaction with strong laser field.

2.4. High-Order Harmonic Generation in Fullerenes Using Few- and Multi-Cycle Pulses of Different Wavelengths

2.4.1. *Introduction*

Finding new approaches to improve the efficiency of HHG of laser radiation towards the XUV range is an important goal of nonlinear optics and laser physics. Small sized nanostructures are an attractive approach, since they exhibit local-field-induced enhancement of the nonlinear optical response of the medium [22, 35, 40]. Another mechanism that can enhance harmonic efficiency of clusters is the increase of the recombination cross section of accelerated electron and parent particle in the final step of the three-step mechanism of HHG [20]. It was shown in [47] that laser-irradiated cluster-contained plasmas can emit low-order harmonics efficiently. The advantages of highly efficient harmonic generation in laser-produced plasmas containing nanoparticles have been analyzed in [40].

Fullerenes are an attractive nonlinear medium for the HHG. Their relatively large sizes and broadband surface plasmon resonance (SPR) in the XUV range allowed the first demonstration of enhanced HHG near the SPR of C_{60} ($\lambda_{\mathrm{SPR}} \approx 60\,\mathrm{nm}$, with $10\,\mathrm{nm}$ full width at half maximum) [46]. Theoretical studies of HHG from C_{60} using multi-cycle pulses include an extension of three-step model [48], analysis of the electron bound within a thin shell of a rigid spherical surface, with geometrical parameters similar to those of the C_{60} [49], and application of dynamical simulations [50]. Those studies reveal how HHG can be used to probe the electronic and molecular structure of C_{60}. Theoretical investigation of such systems is hampered by the fact that the Hamiltonian of HHG is time dependent and the systems consist of many electrons. The investigation of the influence of the electrons on the resonant HHG can be performed by means of a multiconfigurational time-dependent Hartree-Fock (MCTDHF) approach, which has the accuracy of the direct numerical solution of the Schrödinger equation and is not significantly more complicated than the ordinary time-dependent Hartree-Fock approach [51]. In particular, the computations could be based on the multiconfigurational time-dependent Hartree software packages. In [52], simulations of resonant HHG were performed by means of a MCTDHF approach for three-dimensional fullerene-like systems. The influence of the SPR of C_{60} on the harmonic efficiency in the range of 60 nm (E=20 eV) was analyzed and showed the role of resonant effects in the HHG enhancement.

The ionization saturation intensities of different charge states of C_{60} are higher compared to isolated atoms of similar ionization potential [53]. With this perspective, it is interesting to analyze HHG from fullerene molecules in the field of few-cycle laser pulses and compare these studies with those carried out using multi-cycle pulses. The motivation for the refereed work [54] was to analyze the conditions for efficient HHG from plasmas containing C_{60}, using picosecond laser pulses to ablate the fullerene-containing target at high pulse repetition rate (1 kHz) and then few-cycle laser pulses ($\tau = 3.5$ fs) to generate the harmonics in the fullerene plasma. HHG in fullerenes was also studied using 1300 nm radiation and compared with those using 780 nm multi-cycle pulses. A theoretical study of fullerene plasma HHG was carried out.

2.4.2. Results and discussion

As in two previous cases, the same scheme of experiments was used in these studies (see Fig. 2.1). Two types of targets were studied: (i) C_{60} powder, which was glued onto a glass substrate or onto a rotating aluminum rod, and (ii) bulk graphite for plasma harmonic generation. The maximum harmonic conversion efficiency was obtained by varying the distance between the femtosecond beam and the target, as well as the z-position of the plasma. The harmonic spectra from plasmas produced on the bulk graphite and fullerene powder glued onto the glass surface are presented in Fig. 2.10a in the case of laser wavelength 780 nm. Harmonics up to the 29th order

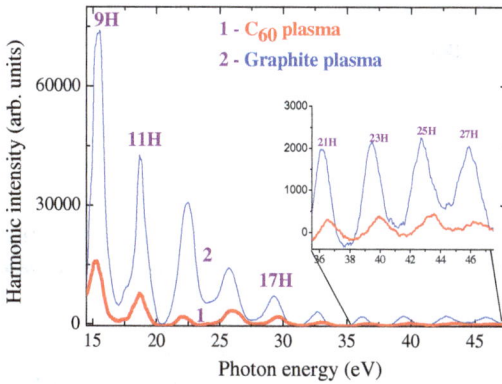

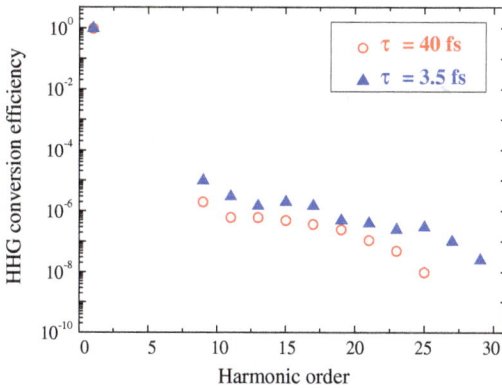

Fig. 2.10. (a) Harmonic spectra from fullerene (thick curve) and graphite (thin curve) plasmas using the heating 8 ps pulses and 3.5 fs probe pulses under identical experimental conditions. (b) Comparison of HHG conversion efficiencies from fullerene plasma using the 3.5 fs (triangles) and 40 fs (circles) probe pulses. Here the laser wavelength is 780 nm. Reproduced from [54] with permission from Optical Society of America.

were obtained from the fullerene plasma. Note that the harmonic efficiency in the case of the graphite plasma was about five times higher compared with the case of fullerene plasmas, most likely due to the higher concentration of emitters in the former case, as has also been reported in [16].

The results of comparative studies of the HHG in fullerene plasma using few-cycle (3.5 fs) and multi-cycle (40 fs) pulses of 780 nm radiation are presented in Fig. 2.10b. In the latter case, a 1 kHz repetition rate Ti:sapphire laser generating 4 mJ, 40 fs pulses was used. The HHG conversion efficiency for the 40 fs pulses at the beginning of plateau range in the case of the plasma plume containing fullerenes was estimated to be $\sim 5 \times 10^{-6}$ using a comparison with the HHG conversion efficiency in a silver plasma, which has previously been reported at similar experimental conditions to be 1×10^{-5} [40]. One can note that the cutoff in the case of longer pulses (25th harmonic) was shorter with regard to the few-cycle pulses at similar intensities of these pulses in the plasma plume.

The problem with using a fullerene powder-containing target is the shot-to-shot instability and rapid decrease of the harmonic yield, due to the abrupt change in the target morphology following ablation. Such instabilities limit applications of this radiation source and greatly hamper efforts to measure the pulse duration of the harmonic emission, which typically requires a large number of laser shots. A solution to this problem might be the use of long homogeneous tapes containing fullerenes, which continuously move from shot to shot to provide a fresh surface for each next laser pulse. The use of rotating targets containing C_{60} is another method for improvement of the harmonic stability from this medium, which was implemented in this work, similarly to the experiments with uracil powder. The density of fullerenes on the rotating target should be the same as the density of powder, since the fullerene powder was glued without additional pressing. The thickness of powder layer was 2 mm. The uniformity was maintained with accuracy better that 0.1 mm.

Due to small concentration of fullerenes in the plasma plume, the amount of ablated material was insignificant even at 1 kHz pulse repetition rate. So the main difference between the rotating and non-rotating targets was the thermal conditions of heating spot. In the case of fixed target, the melting bath appeared after 1000 shots (e.g. 1 sec of ablation), which considerably worsened the process of plasma formation. Once the target started to move, the previously heated area cooled down and again could be used for efficient ablation and harmonic generation. We observed this phenomenon for many targets. So here we were able to improve the stability of harmonics (and correspondingly stability of plasma formation conditions) not just by using the fresh (non-ablated) surface but by changing the conditions of overheating of the same spot of the target.

Conversion efficiency studies in these two plasmas showed advantages of HHG in the case of graphite plasmas compared with fullerene plasmas [16]. This can be explained by the higher particle density in the graphite plasma. The concentration of fullerenes is below 10^{17} cm^{-3} [46], while the density in the refereed studies was estimated for carbon plasmas based on a three-dimensional molecular dynamical simulation of laser ablation of graphite using the molecular dynamics code ITAP IMD [55] and showed that for heating by 8 ps laser pulses the graphite plasma density can reach 2.6×10^{17} cm^{-3} at the moderate ablation intensity of 2×10^{10} W cm^{-2}. Another reason for the observed superior features of graphite plasma harmonics could be the production of clusters during laser ablation, though their involvement in HHG requires additional studies, including time-of-flight measurements.

Figure 2.11 shows the comparison of fullerene harmonic spectra generated in the case of 1300 and 780 nm multi-cycle probe pulses. Harmonics up to the 41st order (Fig. 2.11, bottom panel) were observed in the case of 1300 nm probe pulses at the conditions of optimal plasma formation using the heating 160 ps pulses. The application of less intense, longer wavelength (1400 nm) pulses available by tuning the OPA did not result in an extension of harmonic cut-off compared with the case of probe 1300 nm pulses. This observation suggests that the harmonic generation occurred under saturated conditions, with the expectation of stronger harmonics once the micro- and macro-processes governing frequency conversion are optimized. Over-excitation of target by 160 ps pulses (Fig. 2.11, middle panel) led to appearance of plasma emission in the 25–45 eV range of photon energies ($\lambda = 27-50$ nm). Under these conditions, no harmonics were observed during propagation of femtosecond pulses through such over-excited plasma.

The harmonic spectrum up to the 25th order in the case of 780 nm, 40 fs probe pulses is presented in the Fig. 2.11 (upper panel). One can clearly see the extension of harmonic cut-off (from the point

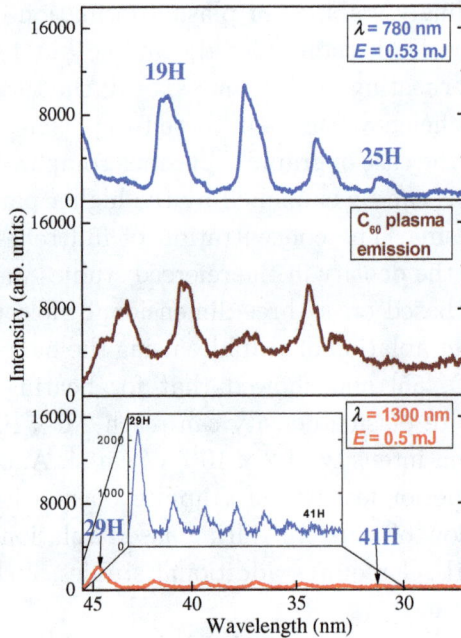

Fig. 2.11. Comparative harmonic spectra from fullerene-contained plume using the 780 nm (upper panel) and 1300 nm (bottom panel) multi-cycle pulses (intensity of heated 160 ps pulses $I_{ps} = 1 \times 10^{10}$ W cm^{-2}). The intensity axes are on the same scale allowing a direct comparison between the three cases. The middle panel shows C_{60} plasma emission spectrum at over-excitation of target by 20 ps pulses ($I_{ps} = 4 \times 10^{10}$ W cm^{-2}), without further excitation by femtosecond probe pulses. Reproduced from [54] with permission from Optical Society of America.

of view of highest harmonic order) in the case of longer-wavelength probe pulses by comparing the HHG spectra using the 780 and 1300 nm probe pulses, while the extension of cut-off energy was insignificant.

Approximately equal energies of driving pulses were maintained in these cases (0.53 mJ for 780 nm pulses and 0.5 mJ for 1300 nm pulses). The intensity of 780 nm radiation in the plasma area was calculated to be $I_{fs} = 4 \times 10^{14}$ W cm^{-2}. The pulse durations of these sources (780 and 1300 nm) were approximately the same (40 and 35 fs respectively). Probably due to phase modulation and propagation through multiple optical elements in optical parametric oscillator-amplifier, the diameter of the 1300 nm beam in the plasma plume was bigger than in the case of 780 nm beam. The corresponding lower intensity of 1300 nm radiation could be responsible for less expected extension of harmonic cut-off energy. From the cut-off formula, we could expect the generation of harmonics up to the 47th order (in the case of 780 nm radiation), well above the observed cut-off (25th harmonic), which probably points out the difference in expected and actual intensity in the plasma area at the optimal conditions of HHG. The reason for this discrepancy could be related with the self-defocusing of driving pulses in the medium containing free electrons. Below we also discuss the cut-off law and its limit of validity at the used experimental conditions.

In atoms the maximum emitted photon energy (giving the cut-off position) is described by the well known relation $E_M = I + 3.17U_p$ with I the ionization energy and U_p the ponderomotive energy. By invoking the three step model this formula can be derived (by use of mere energy conservation and Newton laws) under the assumptions that the laser field amplitude is constant. Rapidly varying laser pulses can considerably shift the cut-off. The molecules support more returning trajectories than atoms and allow the existence of several plateau [56]; it is not therefore easy to disentangle the effects of pulse shape and presence of a molecule and to state a general law for the position of the cutoff; the $E_M = I + 3.17U_p$ law can therefore be used only as a touchstone.

It was observed that the plasma harmonic yields from the 780 and MIR probe pulses are consistent with the predicted single-atom harmonic intensity wavelength scaling $I_h \propto \lambda^{-5}$, which arises due to electron wavepacket spreading before recollision. The harmonic efficiency of the XUV radiation in the range of 30–50 nm driven by MIR pulses was 7 to 15 times less compared with the case of 780 nm probe pulses, which is comparable with the expected ratio between harmonic intensities from these sources ($(1300/780)^5 \approx 12.7$) followed from above rule, assuming approximately equal energies of the 780 and 1300 nm pulses (0.53 and 0.5 mJ respectively).

The fact that HHG in the fullerene system exhibits a wavelength scaling that is consistent with single-atom predictions encouraged us to employ a theoretical model [57] in which the interaction of short pulses with C_{60} was treated in the single active electron approximation with one electron constrained over a structureless spherical surface of radius $R = 3.55 \times 10^{-8}\,\mathrm{cm} = 6.71a_0$. This simplified model can be treated in an analytical fashion. It permits a physical understanding of the dynamics of the active electron, and allows an easy check of resonances between bare molecular states. Indeed a generalization of the model can be used to analyze the physical properties, the energy content and stability of hollow spherical clusters [58]. The calculated harmonic spectra from C_{60} are presented in Fig. 2.12 in the case of 780 nm (photon energy $E_{ph} = 1.6\,\mathrm{eV}$) and 1300 nm (photon energy $E_{ph} = 0.96\,\mathrm{eV}$) pulses propagating through the fullerene medium. The calculations were carried out for 2-cycle pulses ($t = 5.2\,\mathrm{fs}$) and 12-cycle pulses ($t = 31\,\mathrm{fs}$) of 780 nm radiation and 8-cycle pulses ($t = 34\,\mathrm{fs}$) of 1300 nm radiation and intensity $6 \times 10^{14}\,\mathrm{W}\,\mathrm{cm}^{-2}$, which were close to the conditions of fullerene HHG experiments.

The spectra are formed of well-resolved harmonics but with broadened lines (in the case of short pulses) and hyper Raman lines (in the case of long pulses). Hyper Raman, lines with frequency other than harmonics, are due to transitions between laser dressed molecular states [59]. The presence of these lines has been predicted since the very beginning of the theoretical treatment of HHG [60] and found in different contexts such as two-level approximation, quantum dots calculations, hydrogen atom and so on [61–64] but never observed in actual experiments. Several explanations have been proposed to explain this failure; for example it has been argued that they add destructively in the forward direction or that they show a transient nature and are thus overwhelmed by the presented odd harmonics [65, 66].

The calculations showed well-defined harmonics (up to $H_c = 31$) in the case of 1300 nm multi-cycle pulses. The theoretical model exploited the spherical symmetry of the C_{60} by introducing a number of approximations, the most important of which is that the molecule cannot be ionized. This approximation deserves some comment. In spite of the ionization suppression of the C_{60} molecule,

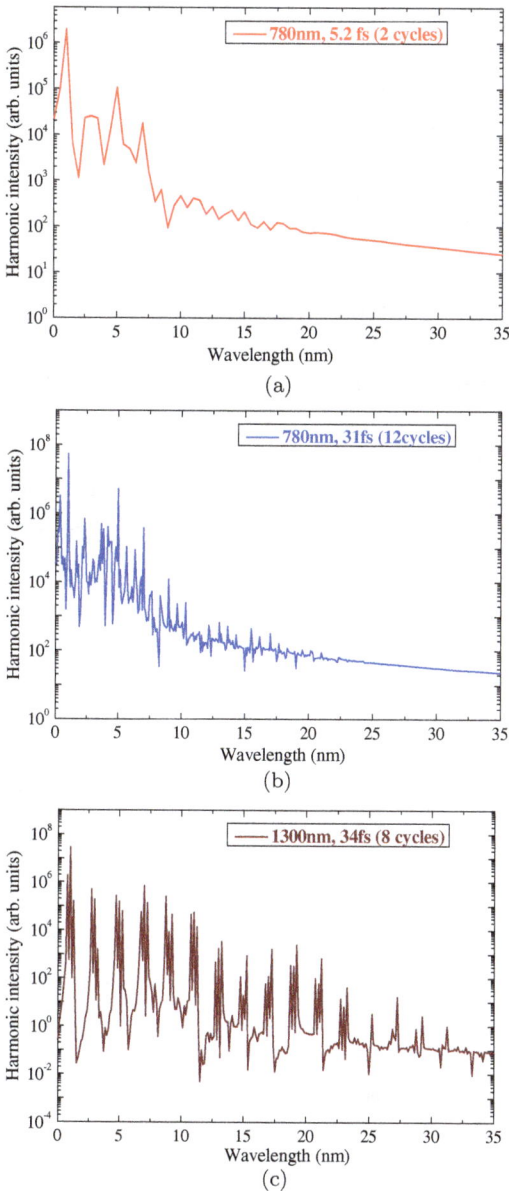

Fig. 2.12. Calculated data of the harmonic spectra from fullerene plasma in the case of (a,b) 780 nm and (c) 1300 nm probe radiation. The pulse durations are (a) 2 optical cycles (5.2 fs), (b) 12 optical cycles (31 fs) and (c) 8 optical cycles (34 fs). Reproduced from [54] with permission from Optical Society of America.

some ionization is bound to occur so that the theory becomes unreliable when ionization becomes significant. The harmonic cut-off in the case of 1300 nm radiation was extended compared with 780 nm radiation ($H_c = 17$), analogously as in the case of experiment.

Previous HHG studies in fullerenes were performed using the multi-cycle pulses (30 fs, 48 fs, and 110 fs [40]). The stability of C_{60} against fragmentation in multi-cycle laser fields leads to fast diffusion of the excitation energy. Even better conditions can occur in the case of few-cycle pulses used for fullerene HHG. In that case fullerenes can withstand the influence of the strong field of few-cycle pulses due to the increase of the ionization saturation intensity as the pulse duration is decreased. This can further increase the diffusion of the excitation energy within the fullerenes due to their very large number of internal degrees of freedom. The increase of energy diffusion is expected to lead to a decrease in the disintegration of fullerenes, which enhances the probability of harmonic emission from these molecules.

The graphite plasmas, which were compared in our experiments with C_{60}, can easily be aggregated during laser ablation, thus leading to nanoparticles in the laser plume. In that case the comparison of two clustered species (large 5–20 nm carbon clusters and 0.7 nm C_{60}) can lead to their different nonlinear optical response once the interacting laser pulse becomes compressed from the multi-cycle to few-cycle duration. The presented results show that for the shortest laser pulses, the HHG cutoff for fullerene is extended, while the harmonic intensity is reduced compared with 'carbon harmonics'. The intensity of the probe femtosecond pulse is an important parameter for optimizing the HHG from C_{60}. Increasing the intensity of the probe pulse did not lead to an extension of the harmonic cutoff from the fullerene plume, which is a signature of HHG saturation in the medium. Moreover, at relatively high probe intensities, a decrease in the harmonic output was observed, which can be attributed to the phase mismatch caused by propagation effects.

The stability of C_{60} molecules against ionization and fragmentation is of particular importance, especially for their application as a medium for HHG using laser pulses of different duration.

The fullerene molecules ablated off the surface should be intact when the probe pulse arrives. Hence, the heating pulse intensity also becomes a sensitive parameter. At lower intensities the concentration of the clusters in the plume would be low, while at higher intensities one can expect fragmentation. This phenomenon is observed when the heating pulse intensity on the surface of fullerene-rich targets is increased above the critical value (Fig. 2.11, middle panel). The abrupt reduction in harmonic intensity in that case can be attributed to phenomena such as fragmentation of fullerenes, an increase in free electron density, and self-defocusing of the probe laser, all of which are expected to reduce the efficiency of HHG.

2.5. Conclusions to Chapter 2

In this Chapter, the application of MIR pulses at high pulse repetition rate for harmonic generation in laser-produced plasmas was presented.

We discussed harmonic generation studies carried out using ablation carbon plasma as a nonlinear medium. We systematically analyzed plasma HHG using ultrashort (3.5 and 40 fs) laser pulses. We showed the efficient frequency conversion of 3.5 fs laser radiation in the range of 15–26 eV using optimally prepared plasma plumes and demonstrated the tuning of harmonic spectra by changing the spectro-temporal characteristics of the probe pulse via pressure tuning of the hollow fiber pulse compression stage. We also showed the results of harmonic generation in carbon plasma using 1300 nm, 40 fs pulses, which allowed the extension of the harmonic cut-off while maintaining a comparable conversion efficiency to 780 nm probe radiation, in surprising disagreement with the expected $I_h \propto \lambda^{-5}$ scaling. Strong harmonics from carbon plasma were attributed to the specific properties of this medium, in particular, the presence of clusters in the plume and their involvement in the enhancement of harmonic yield. We presented comparative studies of optical emission spectroscopy of the graphite ablation plume in the visible, UV and XUV ranges, of time-of-flight mass-spectroscopy of plasma components, and of scanning electron microscopy analysis of plasma debris at optimal condition of harmonic generation, from which we concluded the presence of small carbon clusters (C_{10} – C_{30}) in the plume at the moment of femtosecond pulse propagation, which further aggregate after deposition on a nearby substrate. A carbon plasma mass spectrum was discussed, which showed the value of this method for ascertaining the presence of carbon clusters in the plasma during HHG. We presented results of calculations of plasma concentration at different excitation conditions of the ablating graphite target. Our studies revealed the correlation between the appearance of carbon clusters and the strong harmonic yield from this medium and showed the advanced properties of carbon plasma for efficient HHG.

Further, we have presented the studies of the HHG of laser radiation from laser ablation plumes of the RNA nucleobase uracil. Harmonics were observed using both 30 fs, 780 nm and 1300 nm fundamental laser radiation in plasmas generated upon ablation of solid targets with 1064 nm, 10 ns and 780 nm, 160 ps pulses. The intensity of harmonic emission disobeys the wavelength scaling rule and a higher harmonic conversion efficiency than that expected was observed for the mid-infrared driving beam. Contrary to uracil, HHG was not observed in ablation plumes of its counterpart DNA nucleobase thymine under any of the explored conditions. This considerable variance in the nonlinear optical response is related to differences in chemical structure of the two pyrimidine nucleobases. The fragmentation patterns induced by laser ablation in each compound are closely related with the stability of the molecular ring, which is higher in the case of thymine, and give rise to a characteristic distribution of plasma species in the ablation plumes. The uracil plume is richer in small carbon based fragments, which tend to aggregate and constitute highly efficient nonlinear species. On the other hand, the plume of thymine contains a higher density of parent molecules or large ring-based species which are poor harmonic emitters. These studies constitute the first attempt to analyse differences in structural molecular properties through plasma ablation-induced HHG spectroscopy and show the potential of the technique for structural analysis of medium-size organic molecules of biological interest.

Finally, we presented the results of the experimental and theoretical studies of high-order harmonic generation in the plasmas containing fullerenes (up to the 29th and 41st orders for the 780 and 1300 nm probe radiation) under different plasma conditions and laser parameters in the case of few- and multi-cycle pulses. The comparison of harmonics from C_{60}-rich plasmas and plasmas produced on graphite surface showed stronger harmonic yield in the latter case, which was attributed to higher density of the plasma plume and the formation of clusters during the ablation of graphite targets. These experiments with fullerene powder glued onto aluminum rotating rods demonstrated a dramatic improvement in harmonic generation stability compared with static fullerene-containing targets. The comparative studies using 3.5 and 40 fs pulses showed that, for few-cycle pulses, the harmonic cutoff is extended compared with multi-cycle pulses, which can be attributed to reduced fragmentation of C_{60} for the shorter pulse. The comparison of fullerene harmonic spectra generated in the case of 1300 and 780 nm multi-cycle probe pulses showed the extension of generating harmonic orders in the former case. The plasma harmonic yield approximately follows the $I_h \propto \lambda^{-5}$ wavelength scaling rule predicted for single atoms. The maximal conversion efficiency obtained in these fullerene HHG studies was estimated to be 5×10^{-6}. Theoretical calculations of fullerene harmonic spectra were carried out in the single active electron approximation, which showed harmonics up to $H_c = 31$ in the case of 1300 nm multi-cycle pulses and up to $H_c = 17$ in the case of 780 nm multi-cycle pulses.

References

[1] R. Haight, *Appl. Opt.* **35**, 6445 (1996).

[2] S. E. Harris, J. J. Macklin, and T. W. Hänsch, *Opt. Commun.* **100**, 487 (1993).

[3] G. Farkas and C. Toth, *Phys. Lett. A* **168**, 447 (1992).

[4] B. Sheehy, J. D. D. Martin, L. F. DiMauro, P. Agostini, K. J. Schafer, M. B. Gaarde, and K. C. Kulander, *Phys. Rev. Lett.* **83**, 5270 (1999).

88 *Interaction of MIR Radiation and Plasma*

[5] Y. Akiyama, K. Midorikawa, Y. Matsunawa, Y.Nagata, M. Obara, H. Tashiro, and K. Toyoda, *Phys. Rev. Lett.* **69**, 2176 (1992).

[6] K. Krushelnick, W. Tighe, and S. Suckewer, *J. Opt. Soc. Am. B* **14**, 1687 (1997).

[7] S. M. Gladkov and N. I. Koroteev, *Sov. Phys. Usp.* **33**, 554 (1990).

[8] R. Ganeev, M. Suzuki, M. Baba, H. Kuroda, and T. Ozaki, *Opt. Lett.* **30**, 768 (2005).

[9] R. A. Ganeev, High-Order Harmonic Generation in Laser Plasma Plumes, Imperial College Press (2012).

[10] R. A. Ganeev, Plasma Harmonics, Pan Stanford Publishing (2014).

[11] R. A. Ganeev, Frequency Conversion of Ultrashort Pulses in Extended Laser-Produced Plasmas, Springer (2016).

[12] S. Kubodera, Y. Nagata, Y. Akiyama, K. Midorikawa, M. Obara, H. Tashiro, and K. Toyoda, *Phys. Rev. A* **48**, 4576 (1992).

[13] C.-G. Wahlström, S. Borgström, J. Larsson, and S.-G. Petterson, *Phys. Rev. A* **51**, 585 (1995).

[14] R. de Nalda, M. López-Arias, M. Sanz, M. Oujja, and M. Castillejo, *Phys. Chem. Chem. Phys.* **13**, 10755 (2011).

[15] L. B. Elouga Bom, Y. Petrot, V. R. Bhardwaj, and T. Ozaki, *Opt. Express* **19**, 3077 (2011).

[16] Y. Petrot, L. B. Elouga Bom, V. R. Bhardwaj, and T. Ozaki, *Appl. Phys. Lett.* **98**, 101104 (2011).

[17] R. A. Ganeev, T. Witting, C. Hutchison, F. Frank, P. V. Redkin, W. A. Okell, D. Y. Lei, T. Roschuk, S. A. Maier, J. P. Marangos, and J. W. G. Tisch, *Phys. Rev. A* **85**, 015807 (2012).

[18] Y. Pertot, S. Chen, S. D. Khan, L. B. Elouga Bom, T. Ozaki, and Z. Chang, *J. Phys. B: At. Mol. Opt. Phys.* **45**, 074017 (2012).

[19] L. B. Elouga Bom, S. Haessler, O. Gobert, M. Perdrix, F. Lepetit, J.-F. Hergott, B. Carré, T. Ozaki, and P. Salières, *Opt. Express* **19**, 3677 (2011).

[20] P. B. Corkum, *Phys. Rev. Lett.* **71**, 1994 (1993).

[21] R. Torres, T. Siegel, L. Brugnera, I. Procino, J. G. Underwood, C. Altucci, R. Velotta, E. Springate, C. Froud, I. C. E. Turcu, M. Y. Ivanov, O. Smirnova, and J. P. Marangos, *Opt. Express* **18**, 3174 (2010).

[22] C. Vozzi, M. Nisoli, J.-P. Caumes, G. Sansone, S. Stagira, S. De Silvestri, M. Vecchiocattivi, D. Bassi, M. Pascolini, L. Poletto, P. Villoresi, and G. Tondello, *Appl. Phys. Lett.* **86**, 111121 (2005).

[23] J. Tate, T. Auguste, H. G. Muller, P. Salières, P. Agostini, and L. F. DiMauro, *Phys. Rev. Lett.* **98**, 013901 (2007).

[24] K. Schiessl, L. Ishikawa, E. Persson, and J. Burgdörfer, *Phys. Rev. Lett.* **99**, 253903 (2007).

[25] M. Wöstmann, P. V. Redkin, J. Zheng, H. Witte, R. A. Ganeev, and H. Zacharias, *Appl. Phys. B* **120**, 17 (2015).

[26] R. A. Ganeev, C. Hutchison, T. Witting, F. Frank, W. A. Okell, A. Zaïr, S. Weber, P. V. Redkin, D. Y. Lei, T. Roschuk, S. A. Maier, I. López-Quintás, M. Martín, M. Castillejo, J. W. G. Tisch, and J. P. Marangos, *J. Phys. B: At. Mol. Opt. Phys.* **45**, 165402 (2012).

[27] C. Hutchison, R. A. Ganeev, T. Witting, F. Frank, W. A. Okell, J. W. G. Tisch, and J. P. Marangos, *Opt. Lett.* **37**, 2064 (2012).

[28] T. E. Glover, *J. Opt. Soc. Am. B* **20**, 125 (2003).

[29] H. O. Jeschke, M. E. Garsia, and K. H. Bennemann, *Phys. Rev. Lett.* **87**, 015003 (2001).

[30] A. V. Kabashin and M. Meunier, *J. Appl. Phys.* **94**, 7941 (2003).

[31] R. A. Ganeev, U. Chakravarty, P. A. Naik, H. Srivastava, C. Mukherjee, M. K. Tiwari, R. V. Nandedkar, and P. D. Gupta, *Appl. Opt.* **46**, 1205 (2007).

[32] E. A. Rohlfing, *J. Chem. Phys.* **89**, 6103 (1988).

[33] W. R. Creasy and J. T. Brenna, *Chem. Phys.* **126**, 453 (1988).

[34] S. C. O'Brien, J. R. Heath, R. F. Curl, and R. E. Smalley, *J. Chem. Phys.* **88**, 220 (1988).

[35] T. D. Donnelly, T. Ditmire, K. Neuman, M. D. Perry, and R. W. Falcone, *Phys. Rev. Lett.* **76**, 2472 (1996).

[36] J. W. G. Tisch, T. Ditmire, D. J. Frasery, N. Hay, M. B. Mason, E. Springate, J. P. Marangos, and M. H. R. Hutchinson, *J. Phys. B: At. Mol. Opt. Phys.* **30**, L709 (1997).

[37] R. A. Ganeev, Laser *Phys. Lett.* **9**, 175 (2012).

[38] H. Singhal, R. A. Ganeev, P. A. Naik, J. A. Chakera, U. Chakravarty, H. S. Vora, A. K. Srivastava, C. Mukherjee, C. P. Navathe, S. K. Deb, and P. D. Gupta, *Phys. Rev. A* **82**, 043821 (2010).

[39] C. Altucci, R. Bruzzese, C. de Lisio, M. Nisoli, S. Stagira, S. De Silvestri, O. Svelto, A. Boscolo, P. Ceccherini, L. Poletto, G. Tondello, and P. Villoresi, *Phys. Rev. A* **61**, 021801 (2000).

[40] R. A. Ganeev, *J. Modern Opt.* **59**, 409 (2012).

[41] S. Kim, J. Jin, Y.-J. Kim, I.-Y. Park, Y. Kim, and S.-W. Kim, *Nature* **453**, 757 (2008).

[42] A. McPherson, G. Gibson, H. Jara, U. Johann, T. S. Luk, I. A. McIntyre, K. Boyer, and C. K. Rhodes, *J. Opt. Soc. Am. B* **4**, 595 (1987).

[43] R. A. Ganeev, H. Singhal, P. A. Naik, V. Arora, U. Chakravarty, J. A. Chakera, R. A. Khan, P. V. Redkin, M. Raghuramaiah, and P. D. Gupta, *J. Opt. Soc. Am. B* **23**, 2535 (2006).

[44] C. Hutchison, R. A. Ganeev, M. Castillejo, I. Lopez-Quintas, A. Zair, S. J. Weber, F. McGrath, Z. Abdelrahman, M. Oppermann, M Martín, D. Y. Lei, S. A. Maier, J. W. Tisch, and J. P. Marangos, *Phys. Chem. Chem. Phys.* **15**, 12308 (2013).

[45] D. Roca-Sanjuán, M. Rubio, M. Merchán, and L. Serrano-Andrés, *J. Chem. Phys.* **125**, 084302 (2006).

[46] R. A. Ganeev, L. B. Elouga Bom, J. Abdul-Hadi, M. C. H. Wong, J. P. Brichta, V. R. Bhardwaj, and T. Ozaki, *Phys. Rev. Lett.* **102**, 013903 (2009).

[47] S. V. Popruzhenko, D. F. Zaretsky, and D. Bauer, *Laser Phys. Lett.* **5**, 631 (2008).

[48] M. F. Ciappina, A. Becker, and A. Jaron-Becker, *Phys. Rev. A* **76**, 063406 (2007).

[49] M. Ruggenthaler, S. V. Popruzhenko, and D. Bauer, *Phys. Rev. A* **78**, 033413 (2008).

[50] G. P. Zhang, *Phys. Rev. Lett.* **95**, 047401 (2005).

[51] P. V. Redkin and R. A. Ganeev, *Phys. Rev. A* **81**, 063825 (2010).

[52] H.-D. Meyer, U. Manthe, and L. S. Cederbaum, *Chem. Phys. Lett.* **165**, 73 (1990).

[53] V. R. Bhardwaj, P. B. Corkum, and D. M. Rayner, *Phys. Rev. Lett.* **93**, 043001 (2004).

[54] R. A. Ganeev, C. Hutchison, T. Witting, F. Frank, S. Weber, W. A. Okell, E. Fiordilino, D. Cricchio, F. Persico, A. Zaïr, J. W. G. Tisch, and J. P. Marangos, *J. Opt. Soc. Am. B* **30**, 7 (2013).

[55] J. Roth, F. Géahler, and H.-R. Trebin, *J. Modern Phys. C* **11**, 317 (2000).

[56] M. Lein, *Phys. Rev. A* **72**, 053816 (2005).

[57] D. Cricchio, P. P. Corso, E. Fiordilino, G. Orlando, and F. Persico, *J. Phys. B: At. Mol. Opt. Phys.* **42**, 085404 (2009).

[58] D. Cricchio, E. Fiordilino, and F. Persico, *Phys. Rev. A* **86**, 013201 (2012).

[59] N. Moiseyev and M. Lein, *J. Phys. Chem. A* **107**, 7181–7188 (2003).

[60] T. Millack and A. Maquet, *J. Mod. Opt.* **40**, 2161 (1993).

[61] F. I. Gauthey, C. H. Keitel, P. L. Knight, and A. Maquet, *Phys. Rev. A* **52**, 525 (1995).

[62] W. Chu, Y. Xie, S. Duan, N. Yang, W. Zhang, J.-L. Zhu, and X.-G. Zhao, *Phys. Rev. B* **82**, 125301 (2010).

[63] Z.-Y. Zhou and J.-M. Yuan, *Phys. Rev. A* **77**, 063411 (2008).

[64] V. Kapoor and D. Bauer, *Phys. Rev. A* **85**, 023407 (2012).

[65] A. Di Piazza and E. Fiordilino, *Phys. Rev. A* **64**, 013802 (2001).

[66] D. Bandrauk, S. Chelkowski, and H. S. Nguyen, *J. Mol. Struc.* **203**, 735–736 (2005).

Chapter 3

Resonance-induced Enhancement of Harmonics in Metal Plasmas

The tuning of the wavelength of MIR laser sources is one of most important peculiarities of such sources. In this Chapter, we discuss the advantages of the application of tunable mid-infrared pulses for the harmonic generation in the plasma media at the conditions of resonance-enhanced growth of single harmonics. In particular, we analyze the tuning of odd and even high-order harmonics along the resonances of laser-produced plasmas using optical parametric amplifier of white-light continuum radiation (1250–1400 nm) and its second harmonic. We demonstrate the enhancement of tunable harmonics in the regions of 27, 38, and 47 nm using tin, antimony, and chromium plasmas and discuss the theoretical model of this phenomenon. Further, the tuning of harmonics of ultrashort pulses along the strong resonance of indium plasma using tunable MIR radiation also allowed observation of different harmonics enhanced in the vicinity of $4d^{10}5s^2{}^1S_0 \rightarrow 4d^95s^25p^1P_1$ transition of In II ions. We discuss various peculiarities and show the theoretical model of the phenomenon of tunable harmonics enhancement in the region of 62 nm using indium plasma. With the theoretical analysis we can reproduce the experimental observations and characterize the dynamics of the resonant harmonic emissions. Finally, we demonstrate the generation of enhanced tuneable harmonics of the mid-infrared, 65-fs pulses in laser-produced zinc plasma in the regions of Zn II and Z III autoionizing states (77 and 68 nm respectively).

The role of singly and doubly ionized zinc in modification of harmonic spectra was analyzed by variation of plasma formation conditions. Microprocesses and propagation effect are discussed to describe the narrowing of the enhanced emission spectra.

3.1. Resonance Enhancement of Harmonics in Metal Plasmas using Tunable Mid-infrared Pulses

3.1.1. *Introduction*

Resonance-induced enhancement of the harmonics of ultrashort pulses is one of attractive features of coherent radiation frequency conversion in laser-produced plasmas (LPP) [1–3], with enhancement factor of single harmonic approaching 10^2 in the case of indium plasma. Enhanced single harmonics up to the 35th order of Ti:sapphire laser radiation have been reported so far [4]. The use of LPP allowed the analysis of the ionic transitions possessing large oscillator strengths (i.e., high *gf* values, which are the product of the oscillator strength *f* of an atomic transition and the statistical weight *g* of the lower level).

To address the findings reported during those plasma harmonic studies various theoretical approaches were introduced for description of resonant high-order harmonic generation [5–10]. For these approaches it is important that the harmonic wavelength is resonant with the transition between the ground and the autoionizing state (excited state embedded in the continuum) of the generating ion and that this transition possesses strong oscillator strength. Particularly, in the so-called four-step model [7] the ionized and laser-accelerated electron is captured into autoionizing state (AIS) of the parent ion, and, in the final step, the radiative relaxation of this state to the ground state, a harmonic photon is emitted. In this section, we analyze the approach developed in [5] and generalized to the case of a bichromatic laser field with orthogonally polarized components. In this model, the capture into an AIS is replaced by the field-induced excitation of the ground state into this state and the harmonic strength consists of both resonant and nonresonant parts.

The unavailability of tuning of the wavelength of the most frequently used Ti:sapphire lasers significantly restricts the probability of coincidence of the harmonic order and the transition between AIS and ground states possessing large *gf* values. To facilitate the use of plasma harmonic concept for laser-ablation induced HHG spectroscopy one should use the tunable sources of laser radiation allowing the fine tuning of driving pulses and correspondingly harmonic wavelengths along the spectral ranges of strong ionic transitions. Below, we analyze the fine tuning of harmonics in the vicinity of such transitions during HHG using mid-infrared source of ultrashort pulses and its second harmonic in the plasmas produced on the tin, chromium, and antimony targets and analyze the enhancement of those harmonics. We also discuss the theoretical description of observed phenomena [11].

3.1.2. *Experimental conditions for HHG in plasma plumes using tunable MIR pulses*

Experimental setup consisted of a Ti:sapphire laser, traveling-wave optical parametric amplifier of white-light continuum, and high-order harmonic generation scheme using propagation of amplified signal pulse from OPA through the extended LPP. Part of the uncompressed radiation from the Ti:sapphire laser (806 nm wavelength, 5 mJ pulse energy, 350 ps pulse duration) was separated from a whole beam and used as a heating pulse for homogeneous extended plasma formation using the 200 mm focal length cylindrical focusing lens installed in front of the extended solid target placed in the vacuum chamber. The intensity of the heating pulse on the target surface was varied by up to $4 \times 10^9 \, \text{W cm}^{-2}$. The ablation sizes were $5 \times 0.08 \, \text{mm}^2$.

The compressed pulses of Ti:sapphire laser (806 nm, 6 mJ, 64 fs) was operated at 10 Hz repetition rate and pumped the OPA (HE-TOPAS Prime, Light Conversion). Signal and idler pulses from OPA allowed tuning along the 1200–1600 nm and 1600–2600 nm ranges respectively [Fig. 3.1(a)]. In these HHG experiments, the signal pulses, which were 1.5 times stronger than the idler pulses, were converted in to the harmonics. Most of experiments were

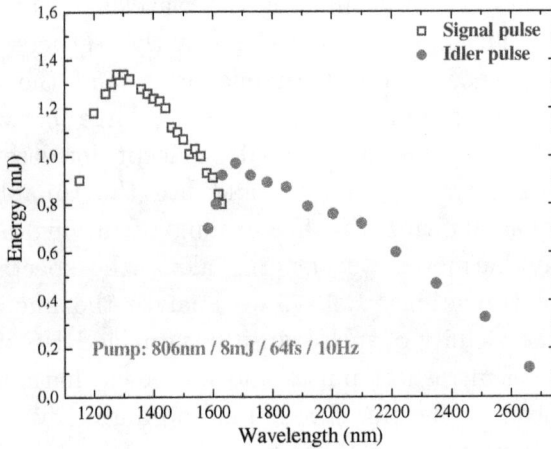

(a)

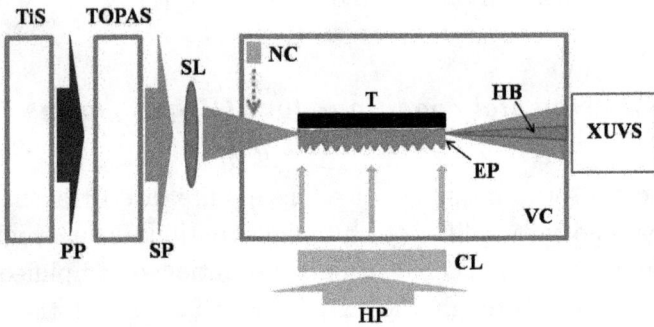

(b)

Fig. 3.1. (a) Spectral tuning of mid-infrared signal and idler pulses from OPA. (b) Experimental setup. TiS, Ti:sapphire laser; TOPAS, optical parametric amplifier; PP, pump pulse for pumping the optical parametric amplifier; SP, amplified signal pulse from OPA; HP, heating picosecond pulse from Ti:sapphire laser; SL, spherical lens; CL, cylindrical lens; VC, vacuum chamber; T, target; EP, extended plasma; NC, nonlinear crystal (BBO); HB, harmonic beam; XUVS, extreme ultraviolet spectrometer. Reproduced from [11] with permission from IOP Publishing.

carried out using the 1 mJ, 70 fs signal pulses tunable in the range of 1250–1400 nm. This variation of driving pulse wavelength was sufficient for tuning the harmonics along various resonances of ionic species. Spectral bandwidth of tunable pulses was 45 nm. The intensity of the 1310 nm pulses focused by 400 mm focal length lens inside the extended plasma was $2 \times 10^{14}\,\text{W cm}^{-2}$. The driving pulses were focused onto the extended plasma from the orthogonal direction, at a distance of $\sim100\mu$m above the target surface [Fig. 3.1(b)]. The plasma and harmonic emissions were analyzed using an extreme ultraviolet spectrometer.

Most of experiments were carried out using the two-color pump of LPP. The reasons for using the double beam configuration to pump the extended plasma is related to the small energy of the driving MIR signal pulse ($\sim1\,$mJ). The $I_H \propto \lambda^{-5}$ rule (I_H is the harmonic intensity and λ is the driving field wavelength) [12] led to a significant decrease of harmonic yield in the case of the longer-wavelength sources compared with 806 nm pump and did not allow the observation of strong harmonics from the $\sim1300\,$nm pulses. Because of this the second-harmonic (H2) generation of signal pulse was used to apply the two-color pump scheme (MIR + H2) for plasma HHG.

Sn, Cr, and Sb plasmas were analyzed as the media for harmonic generation using the tunable source of ultrashort pulses. Previously, those plasmas have shown single harmonic enhancement using the fixed wavelength of pump (Ti:sapphire) lasers. The 5-mm-long samples of above elements were installed in the vacuum chamber for laser ablation.

3.1.3. *Experimental studies of resonance enhancement of MIR-induced harmonics in plasmas*

As it was mentioned, the harmonics generated in above plasmas using MIR pulses (1 mJ, 1300 nm) were significantly weaker compared with the 8 mJ, 806 nm pump due to the λ^{-5} rule. The comparison of these two pumps at similar energies of pulses (1 mJ) showed six-fold growth of the harmonic yield from various plasmas using 806 nm pulses compared with the 1320 nm pulses, while the theoretical prediction of this ratio was $(\lambda_{1320\text{nm}}/\lambda_{806\text{nm}})^5 = 11.8$. The observed harmonic cutoff in the case of MIR pulses was lower compared with the 806 nm pulses, contrary to the theoretically expected extension of the cut-off energy for longer-wavelength pump ($E_{\text{cut-off}} \propto \lambda^2$), due to very small conversion efficiency, which significantly restricted the observation of harmonics below the 50 nm spectral region.

The efficiency of HHG can be significantly increased using a two-color field consisting of the fundamental and second harmonic waves, which are mutually orthogonally polarized [13, 14]. In order to achieve such a field combination the BBO crystal was inserted into the path of the focused driving beam. The 0.5-mm-thick BBO crystal (type I, $\theta = 21°$) was installed inside the vacuum chamber [Fig. 3.1(b)]. The conversion efficiency of 650 nm pulses was ~27%. The crystal was tuned for each specific wavelength of used MIR pulses to generate maximal second harmonic yield. Relatively high conversion efficiencies of the second harmonic was obtained in these studies. In the case of 0.7-mm-thick BBO, 33% conversion efficiencies of the second harmonic was achieved, which could be more preferable for the two-color pump experiments. However, the temporal walkoff of two pulses after leaving the crystal becomes more influential than the growth of second wave intensity. Since the HHG was optimized by different means, the length of BBO crystal was one of these parameters to be taken into account. The spectral bandwidth of second-harmonic pulse was 22 nm. The two orthogonally polarized pump pulses (MIR + H2) were overlapped both temporally and spatially in the extended plasma which led to a significant enhancement of odd harmonics, as well as generation of the even harmonics of similar intensity as the odd ones.

The two-color pump drastically modified the harmonic spectra. The extension of the observed harmonic cutoff, significant growth of the yield of odd harmonics compared with single-color (MIR) pump, comparable harmonic intensities for the odd and even orders along the whole range of generation, and tuning of harmonics allowing the optimization of resonance-induced single harmonic generation were among the advanced features of these two-color experiments. Below we discuss the results of the studies of resonance-enhanced single harmonic generation in various plasmas using the tunable MIR + H2 orthogonally polarized pump pulses. The advantage of the studies of resonance-induced enhancement of harmonics using OPA is the opportunity for fine tuning of this high-order nonlinear optical process for spectral enhancement of the harmonic yield. Below we show some examples of the resonance enhancement of the odd and even harmonics using Sn, Sb, and Cr ablations and the adjustment of different harmonics with regard to the same ionic transition.

Prior to resonance enhancement studies of harmonics using MIR + H2 pulses the above targets were analyzed using the pulses from a conventional Ti:sapphire laser ($\lambda = 806$ nm, $E = 3$ mJ, $I = 6 \times 10^{14}$ W cm^{-2}). Figure 3.2(a) shows the enhancement of the 17th harmonic (H17) in tin plasma, H21 in antimony plasma, and H29 in chromium plasma marked by parallelepipeds. As it has been mentioned, these targets were chosen due to known ionic transitions responsible for those enhancements. Studies of the oscillator strengths of those transitions were reported in [15–19]. This figure shows the comparative studies of the HHG in the three above plasmas. One can see the relative efficiencies obtained from these plasmas by comparing the Y-axes. The highest conversion efficiency was observed in the case of Sb plasma.

The use of tunable broadband MIR radiation allowed the analysis of the enhancement of the groups of harmonics close to the resonances possessing strong oscillator strengths, which caused harmonic enhancement in the case of relatively narrowband 806 nm pulses. Chromium plasma showed the enhancement of harmonics close to the 27 nm region where strong transitions of Cr II significantly modify the featureless decay of plateaulike harmonic spectrum [see the raw images of harmonics obtained using CCD camera, Fig. 3.2(b)]. One can see the growth of harmonic yield in this spectral region for different groups of harmonics during tuning of the wavelength of driving pulses. In particular, harmonics starting from H46 and higher were stronger compared with lower order ones in the case of 1280 nm + H2 pump (upper panel). Similar features were observed for other groups of harmonics in the case of 1310 nm + H2, 1380 nm + H2, 1420 nm + H2, and 1460 nm + H2 pumps (other panels). Dashed lines show the tuning ranges of two harmonics (H20 and H25). In the case of 1420 nm + H2 pump the harmonics were extended above the 60s orders of MIR radiation.

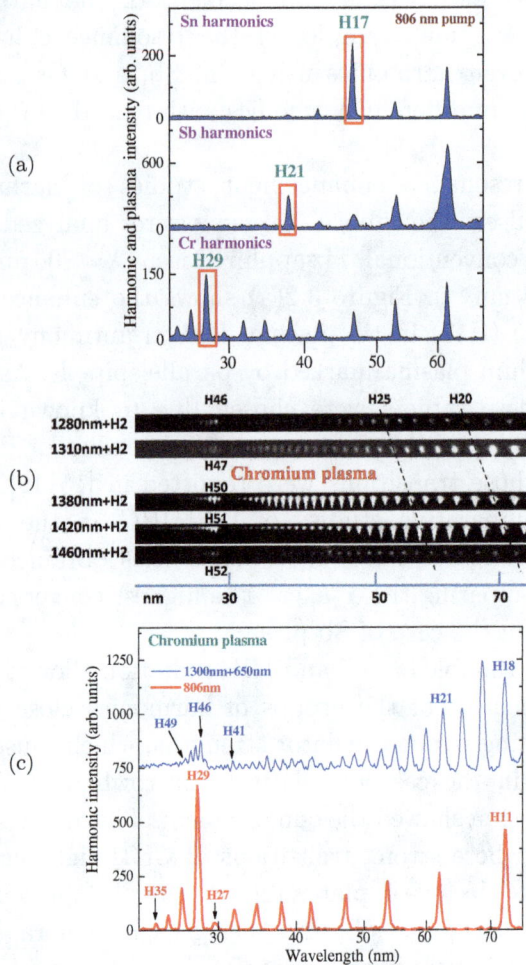

Fig. 3.2. (a) Harmonic spectra using 806 nm driving pulses in Sn, Sb, and Cr plasmas showing the resonance enhancement of the single harmonics (H17, H21, and H29 respectively). (b) Raw images of the tunable harmonic spectra using the pump of chromium plasma by MIR + H2 pulses. MIR pulses were tuned in the range of 1280–1460 nm. Dashed lines in this and other figures show the tuning of specific harmonics. One can see the notable enhancement of harmonics in the vicinity of 27 nm and significant decrease of harmonic yield in the range of 29.5–31 nm. (c) Comparative spectra of the harmonics generated in Cr plasma using the 1300 nm + 650 nm and 806 nm pumps. The MIR-induced curve was shifted along the Y-axis for better visibility and comparison with the 806 nm induced harmonic spectrum. Reproduced from [11] with permission from IOP Publishing.

In above case, we combined three raw images. The purpose in showing raw images is to acquaint the reader with real collection data and visually demonstrate the appearance of the separated group of enhanced harmonics in the case of chromium plasma. The saturated images were chosen intentionally to present the spectra for better viewing. Note that the unsaturated images were used for the line-outs of the HHG spectra shown in other figures. The X-axis is shown in the figure on the basis of the calibration of XUV spectrometer for better viewing of the distribution of harmonics along the short-wavelength region. The HHG spectrometer was calibrated using plasma emission from the used ablated species, as well as other ablating targets. The data on plasma emission of various elements were taken from the NIST Atomic Spectra Database [20].

The enhanced H29 from the ~800-nm-class lasers in the case of chromium plasma was reported in previous studies [21]. In Fig. 3.2(c), we show the comparative line-outs of harmonic spectra in the case of 1300 nm + H2 and 806 nm pumps. In the latter case (bottom curve), one can see a significant suppression of the 27th harmonic ($h\nu = 41.53$ eV) followed with the enhanced H29 ($h\nu = 44.61$ eV). The harmonic spectrum obtained in the case of 1300 nm + 650 nm pump is shown in the upper curve. One can see that the harmonics in these two cases were affected by the same ionic transitions of chromium ions.

The reasons for application of the fixed wavelength source in the above-described case for the HHG in resonance conditions are obvious. Previous experiments, which demonstrated the resonance enhancement of H29 in the chromium plasma using 800-nm-class lasers, did not allow thorough analysis of this process from the point of view of the role of the resonances on the growth of conversion efficiency in the case of a *single* harmonic. The reviewed studies, which use the tunable pump waves, allowed the role of these resonances to be defined. One can see in Fig. 3.2(c) that only H29 (and partially H31) was enhanced, while in the case of 1300 nm + 650 nm pump the maximum enhancement (H46) did not coincide with the wavelength of H29 of 806 nm pump. The wavelength of H29 (27.97 nm) of 806 nm pump is rather positioned between the wavelength of H47 (27.66 nm) of 1300 nm pump [upper panel of Fig. 3.2(c)] and the wavelength of H46 (28.26 nm) of the same pump. Thus it is became obvious that there is some transition corresponding to the group of resonances of Cr II spectra, which lies out of the wavelength of H29 of 806 nm pump and affects the nonlinear optical response of the plasma. This is a clear example of the nonlinear spectroscopy of plasma using high-order processes, which could not be realized in the case of the fixed wavelength sources. That is why HHG from 806 nm and 1390 nm + H2 pumps was compared.

Previous studies of photoabsorption and photoionization spectra of Cr plasma in the range of 41–42 eV [15] have demonstrated the presence of strong transitions, which could be responsible for a suppressed pattern of harmonic spectrum in the wavelength region of 29.5–31 nm. The region of "giant" $3p \rightarrow 3d$ resonances (44–45 eV, $gf = 0.63$) of Cr II spectra was analyzed is [15, 19] and the strong transitions that could enhance the nonlinear optical response of the plume were revealed.

Similar features were observed in the case of harmonic generation in the tin LPP. Studies of the resonant enhancement of HHG in a tin plasma using the 806 nm driving pulses [Fig. 3.2(a), upper panel] showed a strong 17th harmonic analogous to those reported in previous studies of this plasma medium [22,23]. In the present experiment a stronger 17th harmonic was observed, with an enhancement factor of 8× compared with neighboring harmonic orders. The following studies using tunable MIR pulses and their second harmonics have shown a fine tuning of the resonance-enhanced harmonic and change of the order of this harmonic [Fig. 3.3(a)]. The maximally enhanced harmonics, for which both micro-processes and macro-processes were optimized to generate highest photon yield, were changed from H27 in the case of 1290 nm + H2 pump (bottom panel) to H31 in the case of 1450 nm + H2 pump (upper panel). In all these cases, the preceding harmonics were suppressed compared with resonance-enhanced ones (for example, compare H29 and H27 in the case of 1370 nm + H2 pump), though not as strongly as in the case of chromium plasma.

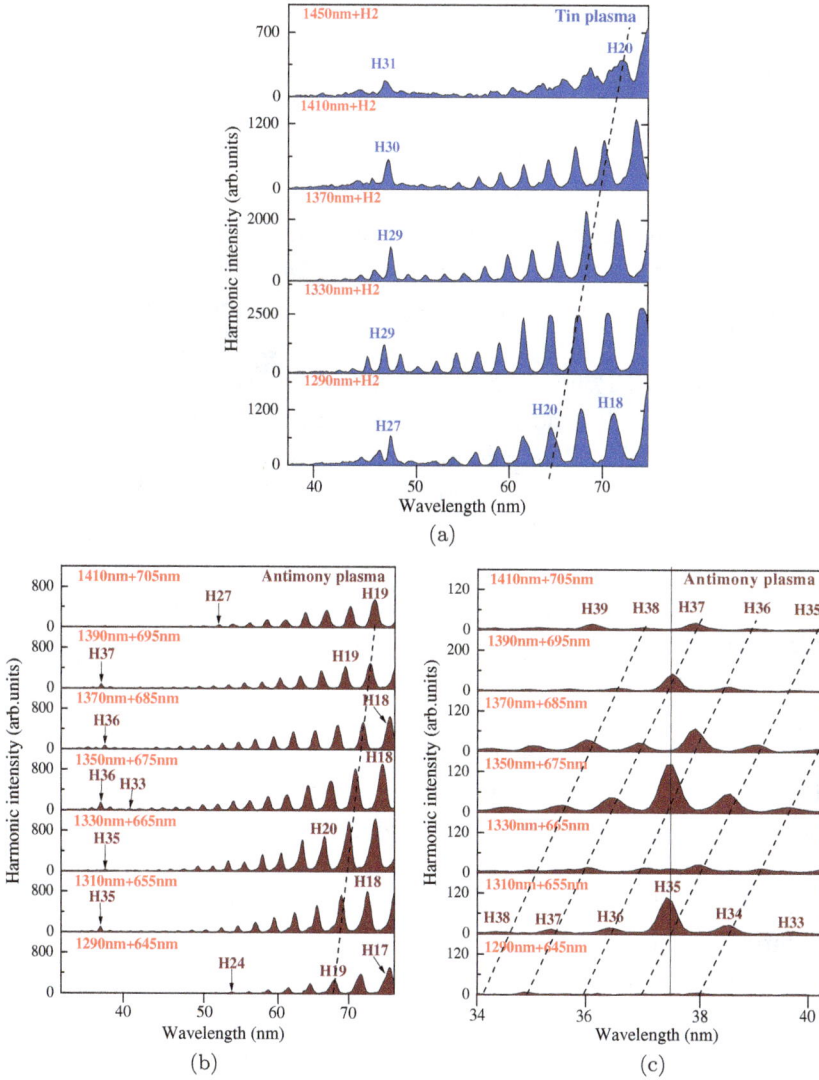

Fig. 3.3. (a) Tuning of resonance-enhanced harmonics generated in the 47 nm region using the two-color pulses propagated through the tin plasma. (b) Harmonic spectra from Sb plasma at different MIR + H2 pumps. (c) Tuning of harmonics in the region of strong ionic transition of antimony (37.5 nm). Reproduced from [11] with permission from IOP Publishing.

One can find from the published data of Sn II transitions in the studied spectral region that the 17th harmonic of 806 nm radiation ($h\nu = 26.15$ eV, $\lambda = 47.41$ nm) is close to some transitions between autoionizing and ground states of tin. The frequencies of these transitions, some of which possess reasonably large oscillator strengths, lie in the photon energy range of 24.9–27.3 eV. The observed enhancement of this harmonic, as well as the H27–H31 of the MIR pulses, were attributed to their closeness with the $4d^{10}5s^25p^2P_{3/2} \rightarrow 4d^95s^25p^2(^1D)^2D_{5/2}$ transition of the Sn II ion ($gf = 1.52$ [18]).

Antimony LPP has previously been analyzed as the nonlinear medium in [20, 24] where single harmonic of Ti:sapphire laser at the wavelength of 37.7 nm was approximately one order of magnitude stronger with regard to the neighboring harmonic orders. Those studies allowed a small tunability of the used Ti:sapphire laser (783–797 nm [24]) to be explored, which led to definition of the maximally enhanced H21 in the case of 791 nm pump. In these studies using 806 nm pulses of Ti:sapphire laser, the 6-fold enhancement of this harmonic ($h\nu = 32.30$ eV) with regard to the neighboring ones [Fig. 3.2(a), second panel] was observed. The application of tunable MIR + H2 pulses allowed the optimal wavelength of pump radiation at which the maximum enhancement of harmonics in the same spectral region was achieved determined [$\lambda = 37.5$ nm, Fig. 3.3(b)]. The tuning of harmonics along this region led to gradual growth and decrease of H35–H37 [Fig. 3.3(c)]. This figure shows the details of the harmonics variations in the narrow spectral range in the vicinity of the resonance transition (37.5 nm) responsible for the enhancement of tuned harmonics.

The resonance enhancement of these harmonics in antimony plasma in the 37–38 nm region is attributed to the strong Sb II transitions [$4d^{10}5s^25p^23P_2 \rightarrow 4d^95s^25p^3(^2D)^3D_3$ and $4d^{10}5s^25p^21D_2 \rightarrow 4d^95s^25p^3(^2D)^3F_3$] at the wavelengths of 37.82 nm (33.78 eV) and 37.55 nm (33.02 eV). The XUV spectra of antimony plasma have been analyzed in [16]. The oscillator strengths of the above transitions have been calculated to be 1.36 and 1.63, respectively, which were a few times larger than those of the neighboring transitions.

The smaller enhancement factor ($\sim 3\times$) of resonance-related harmonics (H35–H37) observed in the present study compared with previous reports ($10\times$ [24], $20\times$ [21]) is attributed to broader pump and harmonic bandwidths, as well as longer wavelength of driving pulses.

Though the above studies were carried out using orthogonally polarized pumps, the interest was to analyze the HHG using parallel polarized pumps as well. For these purposes the idler pulses and 806 nm pump were used. Those studies showed that both orthogonal and parallel polarizations of the two pumps lead to enhancement of the whole harmonic spectrum. In other words, independently of the variation of the spatial trajectory of accelerated electron (i.e., two- and three-dimensional movement) the appearance of even, sum, and difference harmonics, together with odd ones, point out the role of second field as a driving force for broadening the number of coherent frequency components in the XUV spectrum.

Independently of the polarization state of assistant field, the resonance-enhanced mechanism of harmonic amendment remains unchanged. Adding H2 wave, which is orthogonal to the fundamental, one changes the laser-field-driven dynamics of the ionized electron on the microscopic level. In that case, an enhancement of the HHG was observed for atomic harmonics [13], i.e. not only for plasma harmonics.

Resonance enhancement of high-order harmonic generation during the interaction of intense ultrashort laser pulses with various laser ablated plasma plumes has proved to be a promising route towards the production of an intense and coherent XUV radiation source. However, the mechanism of this resonance enhancement is still debated. There are two possible explanations. One relies on a better recombination cross section through an AIS in the single-atom response. The other relies on improved phase matching conditions around the resonance. Some recent findings [25] support the single-atom response hypothesis. The role of the polarization of interacting waves in the case of single-atom response seems insignificant. However, the single-atom response is not fully understood, and further investigation must be carried out.

In the two-color pump HHG experiments in gases, the variation of relative phase between fundamental and second harmonic waves of 30 fs pulses allowed the 3-fold beatings between the long- and short-trajectory induced harmonics to be observed [14]. The use of the focusing optics inside the XUV spectrometer in the reviewed experiments did not allow the influence of the short and long trajectories of accelerated electrons on the divergence of harmonics to be distinguished. However, the difference in the plasma harmonic spectra was observed when the thin (0.15 mm) silica glass plates, which are actually the relative phase modulators, were introduced between BBO crystal and plasma to analyze the variation of the relative phase between two pumps and to compare the change of the relative intensities of "resonant" and "nonresonant" harmonic yields. A significant departure from the large ratio of resonant and nonresonant harmonics in the case of the absence of the glass plates towards the low ratios of these harmonics was observed in the case of propagation through six 0.15-mm-thick plates. Thus the variation of relative phase between pumps may diminish the role of AIS in the single harmonic enhancement.

Below we address the temporal walkoff of two pulses in the nonlinear crystal. In the case of 0.5-mm-long BBO crystal, two pulses were overlapped both temporally and spatially in the extended plasma and allowed a significant enhancement of odd harmonics, as well as generation of even harmonics with the similar intensity as the odd ones.

In the meantime, the temporal walkoff dependent experiments were conducted by using the BBO crystals of different length. The 0.02, 0.3, 0.5, 0.7, and 1.0 mm crystals were used for second harmonic generation [26]. The group velocity dispersion in the BBO crystal leads to a temporal walkoff of two pulses. In particular, due to this effect in the type-I BBO crystal, the 806 nm pulse (ω) was delayed $(\Delta_{\mathrm{cryst}} = d[(n_\omega^o)_{\mathrm{group}}/c - (n_{2\omega}^e)_{\mathrm{group}}/c] \approx 57\,\mathrm{fs}$ for the 0.3-mm-long BBO) with respect to the 403 nm pulse (2ω) due to $n_\omega^o > n_{2\omega}^e$ in this negative uniaxial crystal. Here Δ_{cryst} is the delay between two pulses after leaving the crystal, d is the crystal length, $c/(n_\omega^o)_{\mathrm{group}}$ and $c/(n_{2\omega}^e)_{\mathrm{group}}$ are the group velocities of the ω and 2ω waves in the BBO crystal, c is the light velocity, and n_ω^o and $n_{2\omega}^e$ are the refractive indices of the crystal at the wavelengths of the ω and 2ω pumps. In the case of MIR pulses, the group velocity dispersion between fundamental and second-harmonic waves was notably smaller compared with the case of 806 nm radiation. In the case of 0.5-mm-thick BBO, the calculated walkoff between 1310 and 655 nm waves was 24 fs. The duration of the second harmonic pulse is given by $t_{2\omega} \approx (\Delta_{cryst})^2 + 0.5(t_\omega)^2]^{1/2}$. Hence, the 655 nm beam has longer pulse duration, corresponding to the induced delay and a certain percentage ($\sim$50%) of the fundamental pulse duration. The latter is because the energy of the fundamental radiation is in general not high enough in the leading and trailing parts of pulse to effectively generate the second-order harmonic. $t_{2\omega}$ at the output of the 0.5-mm-long BBO crystal was estimated to be 76 fs, while the 1310 nm pulse duration was 70 fs.

One can see from these calculations that the 1310 and 655 nm pulses were sufficiently overlapped while entering the extended plasma area. One can assume that partial overlap decreased the ratio between the interacting second harmonic and driving pulses and diminished, to some extent, the influence of the 655 nm wave on the output spectrum of generating harmonics. However, the influence of second wave was notably visible, since the odd and even harmonics of similar intensity were easily observable (Figs. 3.2b, 3.2c, 3.3a, and 3.3b). Thus the temporal overlap was sufficient for the observation of the peculiarities of the two-color pump using the MIR + H2 pulses. Contrary to that, once the longer crystals (0.7 and 1.0 mm) were used, a significant modification of harmonic spectra was observed. A decrease of the influence of second wave on the HHG was observed, though the second-harmonic conversion efficiency was increased. The harmonic spectra consisted mainly of the odd orders, with some weak low-order even harmonics. This decrease of the influence of second field was attributed to the temporal walkoff. Thus the SHG effect on the HHG spectrum was uniquely distinguished from a chirp effect in those experiments. There no spatial walkoff was observed between two beams due to installation of BBO crystal between the input window of vacuum chamber and plasma volume.

Notice that, in the case of 806 nm pump, the second harmonic conversion efficiencies and the temporal overlaps of the driving and second harmonic waves, as well as the pulse durations of 403 nm radiation in the cases of different BBO crystals were analyzed. In the case of above mentioned lengths of crystals allowing the 0.4, 5, 9, 11, and 13% second harmonic conversion efficiencies, the ratios of overlapped pulses inside the plasma were 0.004, 0.03, 0.04, 0.01, and 0.007 respectively. One can see the prevalence of using the 0.3- and 0.5-mm-long BBO, since other crystals allowed the observation of less efficient odd and even harmonics generation due to insignificant overlap inside the plasma plume. Moreover, as it has been mentioned, the duration of the second harmonic pulse increases in the case of longer crystals (65, 72, 105, 140, and 195 fs correspondingly; compare with the 64 fs driving pulse). The decrease of the intensity of second harmonic also diminishes the role of this radiation in the variation of HHG spectra. In the case of MIR pulses, lesser dispersion of BBO crystal has diminished the role of walkoff.

Another note is related to the use of extended plasma. The positive dispersion of extended plasma may, to some extent, diminish the temporal delay between fundamental and second harmonic waves caused by the propagation through the negative crystal.

3.1.4. *Theoretical analysis of resonance-enhanced harmonic spectra from Sn, Sb, and Cr plasmas*

In the case of single-color laser field, we use a time-periodic (period $T = 2\pi/\omega$) linearly polarized laser field, with the electric field vector (in dipole approximation) given by

$$\mathbf{E}_{\mathrm{L}}(t) = E_0 \sin(\omega t + \phi)\,\hat{\mathbf{e}}_x, \tag{3.1}$$

where E_0 is the electric field amplitude, and ϕ is an arbitrary phase. For orthogonally polarized bichromatic laser field the electric field vector lies in the xy plane and is defined by

$$\mathbf{E}_{\mathrm{L}}(t) = E_{\mathrm{L}1} \sin(r\omega t + \phi_r)\,\hat{\mathbf{e}}_x + E_{\mathrm{L}2} \sin(s\omega t + \phi_s)\,\hat{\mathbf{e}}_y, \tag{3.2}$$

where $E_{Lj}(j = 1, 2)$, is the jth electric field vector amplitude, $\hat{e}_x$ and $\hat{e}_y$ are the unit polarization vectors along the x and y axis, respectively, the component frequencies $r\omega$ and $s\omega$ are integer multiples of the same fundamental frequency ω, and ϕ_r and ϕ_s are arbitrary phases. The analysis was restricted to the case of linearly polarized bichromatic field components. A more general case of elliptically polarized field components has been analyzed in [27].

One can assume that the target material is such that there is a high radiative transition probability between the ground state with the energy E_1 and a low-lying state having the energy E_2. If the laser frequency is such that the condition $\Delta\omega = E_2 - E_1 = n_R\omega$, n_R - integer, is fulfilled, then, during the single-state HHG process, a coherent superposition of the ground state and this excited state will be formed. In this case, the strength for emission of a harmonic having the frequency Ω takes the form [5, 28]:

$$\mathbf{D}(\Omega) = a_1^2 \mathbf{D}_{11}(\Omega) + a_2^2 \mathbf{D}_{22}(\Omega) + a_1 a_2 \left[\mathbf{D}_{21}(\Omega) + \mathbf{D}_{12}(\Omega)\right], \quad (3.3)$$

where a_1 and a_2 are the initial amplitudes of the bound states in the superposition of states 1 and 2, and $D_{jj'}(\Omega)$ is the Fourier transform of the time-dependent dipole $d_{jj'}(t)$, which is given in [9]. This simple model, which is explained in the Introduction and will be further discussed in subsection 3.1.5 (see [5] for more details and [29] for application), is able to explain qualitatively the experimental data, as it will be shown below.

Note that due to the definition of the electric field vector, both x and y components of $\mathbf{D}(\Omega)$ contribute to the harmonic intensity which is defined by the relation $\Omega^4 |\mathbf{D}(\Omega)|^2$. The hydrogen-like atom model was used and only the $1s(j = 1)$ and $2p$ ($j = 2$) states were taken into account.

Below the numerical examples of HHG from a coherent superposition of Sn^+, Sb^+, and Cr^+ states for $a_1 = a_2 = 1/2^{0.5}$ are presented. In the case of single-color pump, the laser field intensity was 2×10^{14} W cm^{-2} and $\phi = 0$, while for the case of two-color pump ($r = 1, s = 2$) the laser field intensities are $I_1 = 1.5 \times 10^{14}$ W cm^{-2} and $I_2 = 0.5 \times 10^{14}$ W cm^{-2}.

In Fig. 3.4(a), the harmonic yield as a function of the harmonic order are presented for Sn^+ whose ionization potential is 14.63 eV. For single-color laser field the results are presented by the black solid line with filled circles while the numerical results for two-color laser field are presented by the red solid line with filled squares. In the case of two-color field, both phases are equal to zero. In order to achieve a better visibility, numerical spectra for single-color field are shifted down by two orders of magnitude. In the top (bottom) panel of Fig. 3.4(a) the numerical results for fundamental wavelength 806 nm (1370 nm) are presented. In the case of 806 nm, the resonant harmonic is H17, while in the case of fundamental wavelength of 1370 nm the resonant harmonic is H29, in accordance with the experimental results.

Figure 3.4(b) shows the harmonic yield as a function of the harmonic order for Sb^+. The ionization potential is 16.63 eV, while the other laser parameters are the same as in Fig. 3.4(a). Here the fundamental wavelength is 806 nm and the corresponding resonant harmonic is H21. In the case of two-color field and for the fundamental wavelength of 1350 nm the resonant harmonic is H36 and this is in accordance with the experimental results. In the case of single-color field with fundamental wavelength of 1350 nm, emission of even harmonics is forbidden due to the inversion symmetry. In that case $\Delta\omega$ was slightly changed in order to obtain the adjacent resonant harmonic H35. The resonant harmonic is denoted by corresponding color in the upper right corner of the bottom panel in Fig. 3.4(b).

In Fig. 3.4(c), analogous numerical results are presented for the Cr^+ whose ionization potential is 16.48 eV. The other laser parameters are the same as in Fig. 3.4(a). For the fundamental wavelength of 806 nm the corresponding resonant harmonic is H29. In the case of two-color pump and for the fundamental wavelength of 1280 nm the resonant harmonic is H46 which is similar to the experimental results. From the same above-mentioned reasons as in the case of Sb^+ ions, in the case of single-color field it is not possible to obtain this resonant harmonic. $\Delta\omega$ was also slightly changed in order to obtain the adjacent resonant harmonic H47. The resonant

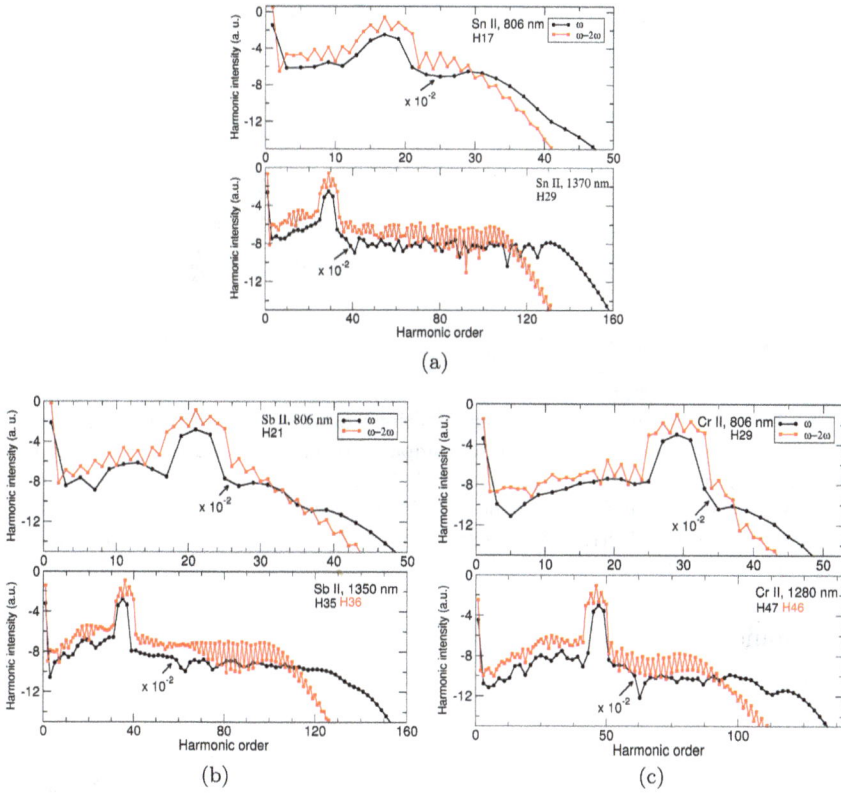

Fig. 3.4. (a) Harmonic intensities as functions of the harmonic order for HHG from Sn II. The laser field intensity is 2×10^{14} W cm^{-2} and $\phi = 0$ for single-color field, and $I_1 = 1.5 \times 10^{14}$ W cm^{-2} and $I_2 = 0.5 \times 10^{14}$ W cm^{-2} with $r = 1$, $s = 2, \phi_r = \phi_s = 0$ for two-color field. The fundamental wavelengths are indicated in the upper right corner of both panels as well as corresponding resonant harmonics. Numerical results for single-color laser field are presented by the black solid line with filled circles while the numerical results for two-color laser field are presented by the red solid line with filled squares. (b) Same as in the Fig. 4(a) but for Sb II. The resonant harmonic for 806 nm is H21, while the resonant harmonic for single-color (two-color) field with 1350 nm is H35 (H36). (c) Same as in the Fig. 4(a) but for Cr II. The resonant harmonic for 806 nm is H29, while the resonant harmonic for single-color (two-color) field with 1280 nm is H47 (H46). Reproduced from [11] with permission from IOP Publishing.

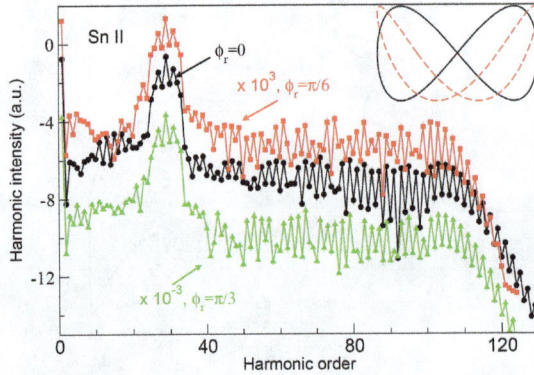

Fig. 3.5. Harmonic intensities as functions of the harmonic order for HHG from Sn II by two-color field whose components have the same intensities and wavelengths as in the lower panel of Fig. 3.4(a), for $\phi_s = 0$ and for three different phases $\phi_r = 0, \pi/6$, and $\pi/3$ as denoted. The inset in the upper right corner shows the corresponding electric field vectors. Reproduced from [11] with permission from IOP Publishing.

harmonic is denoted by the corresponding color in the upper right corner of the bottom panel in Fig. 3.4(c).

In order to show the influence the relative phase between two orthogonal components of the field given by Eq. (3.2), in Fig. 3.5 the numerical results for Sn II are presented. The fundamental wavelength is 1370 nm, $\phi_s = 0$ and the three different phases $\phi_r = 0$, $\pi/6$, and $\pi/3$ are indicated. In order to achieve better visibility, harmonic spectrum for $\phi_r = \pi/6(\phi_r = \pi/3)$ is shifted up (down) by three orders of magnitude. As it can be seen in Fig. 3.5, the cutoff position of the harmonic yield only slightly depends on the relative phases, while the impact on the group of resonant harmonics is negligible. As an inset, in the upper right corner of Fig. 3.5, the electric field vector is shown for phases $\phi_r = 0$ (black solid line) and $\phi_r = \pi/6$ (red dashed line; the field for $\phi_r = \pi/3$ looks the same as the field for $\phi_r = \pi/6$, while the corresponding vector potential $\mathbf{A}_L(t)$, $\mathbf{E}_L(t) = -d\mathbf{A}_L(t)/dt$, changes sign). From these polar plots of the field one can see that there are long time intervals during which the field is close to linear. If, in the first step of the HHG process, the electron appears in the continuum at the beginning of such intervals it is able to return to the parent ion at the end of these intervals and recombine emitting a high harmonics. The situation is similar to the case of HHG by bicircular field [30].

In Fig. 3.6, the harmonic yield is presented as a function of the harmonic wavelength for Sb^+ exposed to the field given by Eq. (3.2). The fundamental wavelength changes from 1290 nm (bottom panel) to 1410 nm with steps of 20 nm. In all presented panels $\Delta\omega = 33.02$ eV was fixed. The corresponding wavelength is denoted in the upper left corner of each panel. As it can be seen in Fig. 3.6, the resonant harmonics H35, H36, and H37 correspond to the fundamental wavelengths 1310 nm, 1350 nm, and 1390 nm, respectively. The intensities of these resonant harmonics decrease when the fundamental wavelength changes in steps 20 nm [see the first, third, fifth and seventh panel of Fig. 3.6 (from bottom to the top)]. One can find a similarity in the behavior of harmonic yield in this calculation and the experiment shown in Fig. 3.3(c).

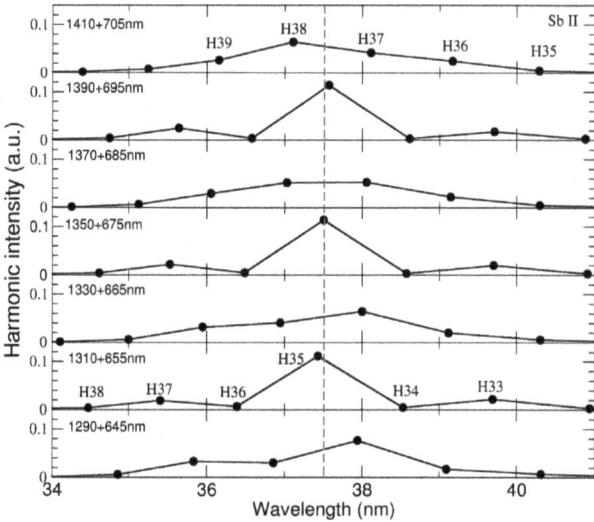

Fig. 3.6. Harmonic intensities as functions of the harmonic order for HHG from Sb II for different fundamental wavelengths starting from 1290 nm (bottom panel) to 1410 nm (top panel). The other laser parameters are as in the lower panel of Fig. 3.4(b). Reproduced from [11] with permission from IOP Publishing.

3.1.5. *Discussion*

Tuning of the pump laser wavelength of Ti:sapphire lasers during resonance harmonic studies has been reported a long time ago [1]. In those studies, an insignificant tuning of driving radiation was accomplished due to lack of the tunability of Ti:sapphire lasers, which is insufficient for observation of the peculiarities of resonance enhancement. In present studies, the parametric waves were used to tune the nearby harmonics along the strong resonances. Under these conditions, the tuning of the wavelength of pump radiation was dramatically larger compared with above-mentioned case, which allowed notable variation of the conditions for resonance enhancement of different orders of harmonics.

The importance of the phenomenon of resonance-induced harmonic enhancement has been proven by numerous recent theoretical papers, which offered various approaches in the explanation of the resonance enhancement of harmonics in LPP. Existing theories of such microprocess as resonance-induced enhancement of single harmonic are mostly based on: (i) the four-step model when the ionized and laser-accelerated electron is captured into single AIS of the parent ion, and, in the final step, the radiative relaxation of this state to the ground state leads to the emission of enhanced harmonic, (ii) the harmonic generation in the presence of a shape resonance using the time-frequency analysis of the intensity and phase, which underlined the resonance enhancement irrespective of the pulse length and supported the four-step model, (iii) the approach in which the capture into an AIS is replaced by the field-induced excitation of the ground state into this state and the harmonic strength consists of both resonant and nonresonant parts, and (iv) usual three-step scenario by applying the factorization formula without additional assumptions.

Among the existing theories of resonance-induced enhancement of single harmonic, the most advanced is the theory of [7], which is based on the four-step model in which the ionized and laser-accelerated electron is captured into single AIS of the parent ion, and, in the final step, the radiative relaxation of this state to the ground state leads to the emission of enhanced harmonic. In the discussed work [11], an earlier developed model [5] in which the capture into an AIS is replaced by the field-induced excitation of the ground state into this state was used. This model is simpler than the four-step model [7], which uses solutions of the 3D TDSE with a model potential. In the approach developed in [11] the harmonic strength consists of both resonant and nonresonant parts. This model has the advantage that the full spectrum with both resonant and nonresonant HHG can be obtained much easier. These two models have not been compared yet. Furthermore, this approach is now generalized to the case of a bichromatic laser field with orthogonally polarized components.

As one can see, plenty of theoretical studies have reported on the possibility of resonance enhancement of harmonics. Most of theoretical studies were aimed on the availability of this process in gas HHG experiments. However, no experimental evidence with gases showing the resonance enhancement of single harmonic has been reported so far. The only study where the partial enhancement of narrow component of single harmonic demonstrated during HHG in Ar gas was related to the influence of the Fano resonances [31].

Why is this the case in gas media, while in plasma media we have some samples of such enhancement? In other words, why have the resonant harmonics not been observed with gas media? Wavelength tunable intense femtosecond lasers are now available, and so there should be no problem tuning the laser wavelength to a specific resonance, whether it be of a plasma or gas. The answer to this fundamental question is related to the basic principles of the role of some resonances in the enhancement of harmonics. Tens of plasma samples have been analyzed, which showed numerous cases of the coincidence of harmonics with some emission lines of corresponding ions. However, only in a few cases has this coincidence resulted in the enhancement of harmonics. In other cases, various factors prevented the observation of resonance enhancement. Among them are the self-absorption near the ionic transitions, which emit strong radiation in the XUV, and weak oscillator strength of these transitions. The self-defocusing in the vicinity of ionic transitions can also be considered as the impeding factor, though its influence was not studied in [11]. However, self-defocusing has clearly been observed and analyzed during past plasma HHG experiments [32].

It is obvious that the choice of a few noble gases is dramatically lower compared with almost all the periodic table containing predominantly solid elements, not to mention thousands of complex solids. Consequently, the probability of finding the appropriate resonance with large oscillator strength, which matches with some harmonics, is significantly smaller in the case of gas HHG compared with plasma HHG. That is why the studies of this phenomenon in plasma media have more chances for success. To add more on that topic we would like to mention that the availability of resonance enhancement largely depends on the population of appropriate energetic levels of ions. These conditions could be more easily realized while producing the laser plasma at "appropriate" conditions of target ablation, rather than in the case of gases.

The pulse durations of both MIR pulses (70 fs) and their second harmonics (75 fs) were measured using standard autocorrelation technique. These data (and correspondingly the data on the intensities of interacting waves) were used for the theoretical modeling of resonance-induced enhancement of harmonics. Note that the pulses were long enough so that in theoretical models an infinitely extended plane wave can be supposed; the used pulses certainly were not the few-cycle pulses for which special technology should be used and for which a modified theoretical model should be applied.

These studies have demonstrated new opportunities in the analysis of the strength of some ionic transitions responsible for enhancement of single harmonic in the plateau region. An example of such an analysis is shown in Fig. 3.3(a), where the intensity ratio of the various harmonics near 47 nm vary as one tunes the pump laser wavelength. It is obvious that the enhancement of single nearby harmonic strongly depends on the coincidence of the harmonic and ionic transition wavelenghts. Correspondingly, the enhancement factor, or intensity ratio between the "resonance" harmonic and neighboring ones, will be obviously changed once the wavelength of enhanced harmonic tunes towards or outwards the transition.

The fact that this process is reproduced for various harmonic orders just confirms the consideration of the fundamental role of the ionic transitions possessing large oscillator strength in the enhancement of harmonics. This phenomenon is qualitatively reproduced in our calculations showing the enhancement of different harmonic orders near the same ionic transition.

Next, in Fig. 3.3(c), the 37.5 nm harmonic is in resonance with the pump laser for three cases, 1390 nm + 695 nm, 1350 nm + 675 nm, and 1310 nm + 655 nm. However, the intensity of the resonant 37.5 nm harmonic varies considerably. The reason for the variation of enhancement factor is also related with different energies of MIR radiation. The largest energy of MIR pulses in these experiments was observed in the case of 1350 nm pulses. We just tuned the OPA and observed the above phenomenon of intensity ratio variations. Meanwhile, Fig. 3.6 presents the simulations for the enhancement of different harmonic orders assuming similar energies of pump radiation.

It was demonstrated that the addition of second field dramatically compensates for the negative consequences of the $I_H \propto \lambda^{-5}$ rule in the case of longer-wavelength source. In particular, in the case of chromium plasma, the use of MIR pulses alone (i.e. without second field) did not lead to harmonic generation, or lead to a few extremely weak harmonics at the longer wavelength region. Thus the expected improvement in cut-off energy for longer wavelength source ($E_{\text{cut-off}} \propto \lambda^2$) did not happen due to much stronger wavelength-dependent decrease of harmonic yield. In Fig. 3.2(c), the MIR + H2 induced HHG curve was shifted along the Y-axis for better visibility and comparison with the bottom curve. One can clearly see the similarity in the harmonic yields in these two cases. Thus the two-color approach in the MIR region led to the cancellation of the disadvantage caused by the above-mentioned wavelength-dependent rule. Secondly, with this curve from 806 nm pump we show the resonance enhancement of single harmonic (H29) in the vicinity of Cr II resonance transition (27 nm). The comparison with MIR + H2 induced harmonics allowed us to conclude about the enhancement of a group of harmonics at the blue side of the resonance transition, while the longer wavelength side showed a significant decrease of the harmonic yield, which could be related with the involvement of propagation effects.

The 806 nm and MIR + H2 induced enhancements of resonance harmonics in chromium plasma are shown in Fig. 3.2c. However, it is difficult to compare the enhancement factors in these two cases. In the case of the former pump, 5 mJ pulses were used, while in the case of the latter pump it was approximately 0.7 mJ + 0.2 mJ energy of those pulses. Even under these unfavorable conditions [i.e., significantly less energy of main (MIR) pulse and above-mentioned wavelength-dependent rule] only a four-fold decrease of the enhancement of resonance harmonic using these pulses was observed. Once 1 mJ pulses of 806 nm pump were used, no harmonics were generated below 40 nm. So from this point of view one can admit the dramatic comparative enhancement of resonant harmonic in the case of used scheme at similar conditions of pump energies.

The resonance enhancement of single high-order harmonics in various laser-produced plasmas has been obtained since the first observation of this process in indium plasma. Tellurium, chromium, manganese, arsenic, selenium, tin, and antimony were among the plasma media where the yield of some harmonics was higher than one from the lower-order harmonics, contrary to the commonly accepted rule of a decrease of conversion efficiency for each next order of harmonics. Those plasma harmonic studies were carried out using the fixed wavelengths of the driving pulses of Ti:sapphire lasers. The lack of tunability did not allow the analysis of the nonlinear optical processes at the conditions of fine tuning of driving pulses for the growth of single harmonic conversion efficiency.

One can speculate in this connection on the availability of using the high energy transitions related to deeper core resonance. It is not obvious that deeper core resonances can decisively influence the HHG. Deeper core resonances may appear for multiple ionized plasma ions (see [33] for inert gas ions). However, the formation of multiple ionized ions would lead to the growth of free electron concentration, which may strongly affect the phase matching condition. In other words the prevalence of the phase mismatch over the single-atom related growth of harmonics due to deeper core resonances may fully cancel the advantages of using the resonance concept.

Though the approaches considering the role of resonances as the main reason for experimentally observed growth of the yield of some single harmonics presently prevail, one has to point out another approach developed in [34]. As it has been shown the energy shifts of atomic states approach to the free atom energy level difference in the case of laser fields of near-atomic strength. Such behavior can be considered from the point of view of atomic field strength definition. It was suggested that the interpretation based on presence of resonances with ionic transitions is doubtful, because in the laser fields of near-atomic strength the motion of atomic electron obeys the action of two equal forces which are due to the intra-atomic and external laser fields.

The above approach using laser dressed states is probably the reason why one cannot take into account the role of resonances (with laser dressing all spectrum is shifted into resonance). But this is just only the theoretical approach developed in the above-mentioned work, contrary to many other approaches and publications, and, in our opinion, it cannot draw general conclusions. These two approaches are extensively discussed in the laser community.

3.2. Indium Plasma in the Single- and Two-color Mid-infrared Fields: Enhancement of Tunable Harmonics

3.2.1. *Introduction*

Searching for new methods for materials science using optical and nonlinear optical approaches is an important goal of laser physics. HHG of laser radiation has long been considered a promising spectroscopic tool to retrieve the structural and dynamical information regarding the nonlinear medium through analysis of the spectra, polarization states, and phase of generated harmonics [35–37]. Photorecombination, the third step in the HHG recollision model

[38, 39], is the inverse process of photoionization [40] and therefore it is expected that HHG and photoionization must exhibit common resonances in the case of both laser-produced plasmas [24, 41] and gases [42].

The resonance peaks in the photoionization and photorecombination cross sections, including autoionizing [43], shape [44], and giant resonances [45] have long been investigated. In contrast, the studies on the role of resonances in HHG are relatively scarce. Forming resonance conditions to enhance the nonlinear optical response of the medium may be an alternative to the phase-matching technique previously used for harmonic enhancement. The role of atomic resonances in increasing the laser radiation conversion efficiency was actively discussed in the framework of perturbation theory at the early stages of the study of low-order harmonic generation [46]. Resonance enhancement introduces a new possibility of increasing the conversion efficiency of a specific harmonic order by more than one order of magnitude. If this effect could be combined with phase-matching effects and/or coherent control of HHG, one could be able to generate a spectrally pure coherent x-ray source with only a single line in the spectrum, much like saturated x-ray lasers produced by ionic population inversions in highly ionized plasmas.

Effects of resonant harmonics using approximately fixed pump laser wavelength have been reported a long time ago [1] using Ti:sapphire lasers. In those studies, small tuning was accomplished either by using chirping technique within pulse bandwidth or by insignificant variation of master oscillator wavelength. In present studies the tuning of pump laser wavelength has been dramatically improved compared with the above-mentioned case, which allowed the intensity ratios of various harmonics particularly near 62 nm in the case of the HHG in the indium plasma to be changed, thus forming the conditions for resonance enhancement of different orders of harmonics near the resonances possessing large oscillator strength. As it has already been mentioned in the previous section, various theoretical approaches for description of resonant HHG were introduced in [5–8, 47–51]. For these approaches it is important that the harmonic wavelength is resonant with the transition between the ground state and the autoionizing state of the generating ion and that this transition possesses strong oscillator strength. Particularly, in the so-called four-step model [7] the ionized and laser-accelerated electron is captured into AIS (i.e., excited state embedded in the continuum) of the parent ion, and, in the final step, during the radiative relaxation of this state to the ground state, a harmonic photon is emitted.

The unavailability of tuning of the wavelength of the most frequently used Ti:sapphire lasers significantly restricts the probability of coincidence of the harmonic order and the transition between AIS and ground states possessing large values of oscillator strength. To facilitate the use of plasma harmonic concept for laser-ablation induced HHG spectroscopy in the XUV range one should use the tunable sources of laser radiation allowing the fine tuning of driving pulses and correspondingly harmonic wavelengths along the spectral ranges of strong ionic transitions.

In the previous section, we raised the question why have similar resonant harmonics not been observed with gas media? Tunable femtosecond lasers are now available, and so there should be no problem tuning the laser wavelength to a specific resonance, whether it be of a plasma or gas. The answer to this question is related to the basic principles of the role of resonances in the enhancement of a nearby harmonics. In most cases the harmonic spectra show the absence of enhancement of the harmonics near strong emission lines. In the meantime, previous studies have demonstrated that the resonance enhancement of a nearby harmonic close to the resonance can be observed in the case of sufficient amount of excited species (i.e., singly or doubly charged ions). Thus the availability of resonance enhancement depends on the population of the appropriate energetic levels of ions.

Studies of high-order nonlinear processes through exploitation of intermediate resonances show that the proximity of the wavelengths of specific harmonic orders and the strong emission lines of ions does not necessarily lead to the growth of the yield of the single harmonic. The nonlinear optical response of the medium during propagation of intense pulses includes, in some particular cases, the resonance-induced enhancement of specific nonlinear optical processes, the absorption of emitted radiation, and the involvement of collective macro-processes, such as the phase-matching between the interacting waves. The mechanism for improvement of the phase-matching conditions for the single harmonic can be interpreted in that case as follows. The refractive index of plasma in the short-wavelength side of some resonant transitions can be decreased due to anomalous dispersion thus allowing the coincidence of the refractive indices of plasma at the wavelengths of the driving and harmonic waves. To analyze this process in depth one has to define the bandwidths of those resonances, relative role of the nonlinear enhancement of harmonic emission and the absorption properties of LPP in the vicinity of resonances, influence of plasma length on the enhancement of the single harmonic, etc.

In this section, we analyze the fine tuning of harmonics in the vicinity of strong In II transition during HHG using the mixture of tunable MIR source of ultrashort pulses and its second harmonic in the LPP produced on the surface of indium target, analyze the enhancement of those harmonics, and compare with single-color MIR pump. We also present the theoretical description of observed phenomena [52].

3.2.2. *Experimental studies of the resonance enhancement of MIR-induced harmonics in the indium plasma*

Experimental setup was similar to the one described in subsection 3.1.2. Indium plasma was analyzed as the medium for harmonic generation using the tunable source of ultrashort pulses. The 5-mm-long sample of bulk indium was installed in the vacuum chamber for laser ablation. The experiments were carried out using both the single-color and two-color pumps of LPP. As was already mentioned in previous chapter, the reasons for using the double beam configuration to pump the extended plasma is related to the small energy of the driving MIR signal pulse ($\sim 1\,\text{mJ}$). The $I_H \propto \lambda^{-5}$ rule (I_H is the harmonic intensity and λ is the driving field wavelength [53]) led to a significant decrease of harmonic yield in the case of longer-wavelength sources compared with the 810 nm pump and did not allow the observation of strong harmonics from the single-color MIR (1310 nm) pulses. Because of this the second-harmonic generation of signal pulse was used to apply the two-color pump scheme (MIR + H2) for plasma HHG. The variation of the relative phase between fundamental and second harmonic waves was analyzed by insertion of the 0.15-mm-thick glass plates in the path of these beams. The plates were introduced between the BBO crystal and LPP. The group velocity dispersion for the driving and second harmonic waves propagating through such plates makes possible the variation of both the relative phase of two pumps and the delay between the envelopes of these pulses.

The conditions of the two-color pump of plasma were also analyzed using different polarizations of interacting waves when the BBO crystal was installed outside the vacuum chamber. The 2-mm-thick calcite plate was installed in front of the BBO crystal under the conditions when the MIR pulses generate H2 in the 0.7-mm-thick BBO placed between the focusing lens and input window of the vacuum chamber. The rotation of calcite affected only the driving pulse by changing the polarization from linear to elliptical and to circular. In that case the two-color pump consisted of the linearly polarized second-harmonic wave and circularly (or elliptically) polarized fundamental MIR pulses.

The harmonics generated in indium plasma using MIR pulses (1 mJ, 1330 nm) were significantly weaker compared with the case of 8 mJ, 810 nm pump due to the above-mentioned wavelength-dependent yield of harmonics. The comparison of these two pumps at similar energies of pulses (1 mJ) showed the seven-fold decrease of harmonic yield in the plateau region using 1330 nm pulses compared with the 810 nm pulses [Fig. 3.7(a)], while the theoretical prediction of this ratio was $I_{1330\,nm}/I_{810\,nm} = (810\,nm/1330\,nm)^{-5} \approx 12$. The observed harmonic cutoff in the case of MIR pulses was approximately similar to the one from the 810 nm pump [27 eV (H29) and 26 eV (H17) respectively], contrary to the theoretically expected extension of the cutoff energy for

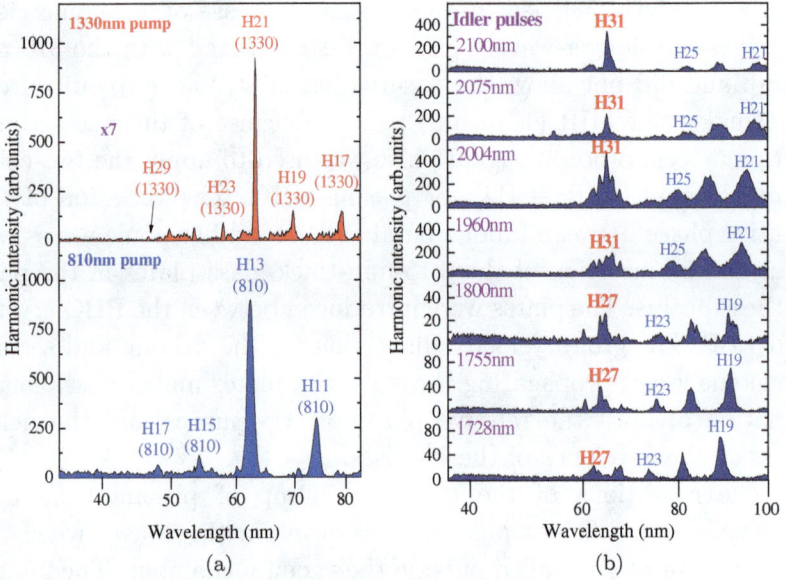

Fig. 3.7. (a) Harmonic spectra using 1330 nm (upper panel) and 810 nm (bottom panel) pulses. Upper panel magnified by a factor of 7 for better comparison of the harmonic spectra generated using different pumps. (b) Spectra of tunable harmonics in the case of single-color pump of indium plasma using idler (1728–2100 nm) pulses of OPA. Reproduced from [52] with permission from American Physical Society.

longer-wavelength pump ($E_{cutoff} \propto \lambda^2$), due to very small conversion efficiency in the case of MIR pulses, which did not allow the observation of harmonics below the 40 nm spectral region. The weak idler pulses from OPA were used as well, with the energy varying in the range of 0.3–0.5 mJ, for harmonic generation in plasma. Even these small energies of driving longer-wavelength pulses were sufficient to observe the resonance enhancement of different harmonic orders near the AIS of indium in the case of the idler pulses tunable along the 1730–2100 nm range [Fig. 3.7(b)].

These studies showed that the use of tunable MIR pulses caused generation of mostly single enhanced harmonic (H21) close to the AIS and a few weak odd harmonics in the longer wavelength range of XUV [Fig. 3.8(a)]. The insertion of the BBO crystal into the path of the focused driving beam drastically modified the harmonic spectra.

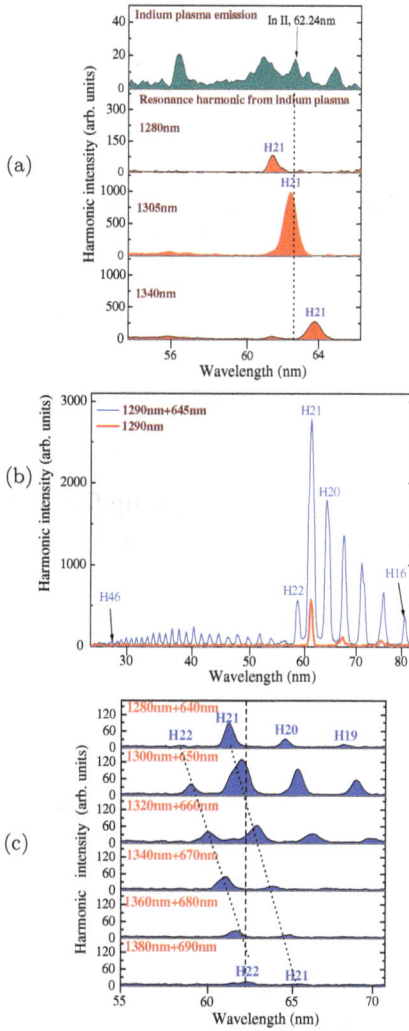

Fig. 3.8. (a) Tuning of H21 along the In II resonance. Upper panel shows the indium plasma emission spectrum. Three bottom panels show resonantly enhanced harmonic using single-color pump (1280, 1305, and 1340 nm respectively). Dotted line shows the position of the resonance transition of In II. (b) Comparative harmonic spectra from indium plasma using single-color (1290 nm; thick curve) and two-color (1290 nm and 645 nm; thin curve) pumps. (c) Relative variations of H21 and H22 yields using different wavelengths of MIR and H2 pump radiation. Dotted lines show the tuning of H21 and H22. Dashed line shows the position of resonance transition. Reproduced from [52] with permission from American Physical Society.

Extension of the observed harmonic cutoff, significant growth of the yield of odd harmonics, comparable harmonic intensities for the odd and even orders along the whole range of generation, tuning of harmonics allowing the optimization of resonance-induced single harmonic generation, as well as a few neighboring orders close to AIS of indium, were among the advanced features of these two-color experiments [Fig. 3.8(b)]. These studies showed that the advantage of the resonance-induced enhancement of harmonics using MIR + H2 pulses is the opportunity of fine tuning of this high-order nonlinear optical process for spectral enhancement of the harmonic yield. Another advantage is closely related with the analysis of the oscillator strengths of some ionic transitions using the HHG approach.

Various schemes of the two-color pump for HHG were introduced in [13, 54–60]. As underlined in [60], a strong harmonic generation in the case of two-color pump is possible due to formation of a quasi-linear field, selection of a short quantum path component, which has a denser electron wave packet, and higher ionization rate compared with the single-color pump. The orthogonally polarized second field also participates in the modification of the trajectory of accelerated electron from being two-dimensional to three-dimensional that may lead to removal of the medium symmetry. With suitable control of the relative phase between the fundamental and second-harmonic pumps, the latter field enhances the short path contribution while diminishing other electron paths, resulting in a clean spectrum of harmonics.

The use of tunable broadband MIR radiation and its second harmonic allowed the analysis of the enhancement of the groups of harmonics close to the resonance possessing strong oscillator strength [Fig. 3.8(c)]. One can see the growth of harmonic yield in the 62.3 nm region of AIS for different groups of harmonics during tuning of the wavelength of driving pulses. These studies using tunable MIR pulses and their second harmonics showed a fine tuning of the resonance-enhanced harmonic and change of the order of this harmonic. The tunability is practically unlimited since the tuning of fundamental wavelength (Fig. 3.1a) allowed shifting the wavelength of high-order

harmonic over the wavelength of neighboring harmonic, which means
the overlap of a full octave.

The top two panels of Fig. 3.9(a) show the harmonic spectra
in the case of using two different schemes when the BBO crystal
was installed either inside or outside the vacuum chamber. In the
latter case, the linearly polarized 1330 nm pulses generated H2 in the
0.7-mm-thick BBO placed outside the vacuum chamber. In the case
shown in the bottom panel of Fig. 3.9(a), the polarization of driving
1330 nm pulses was changed from linear to circular by inserting the
calcite plate in front of BBO crystal. In that case the pump radiation
after propagation of BBO crystal consisted of the circularly polarized
1330 nm pump and linearly polarized 665 nm pump. This apparently

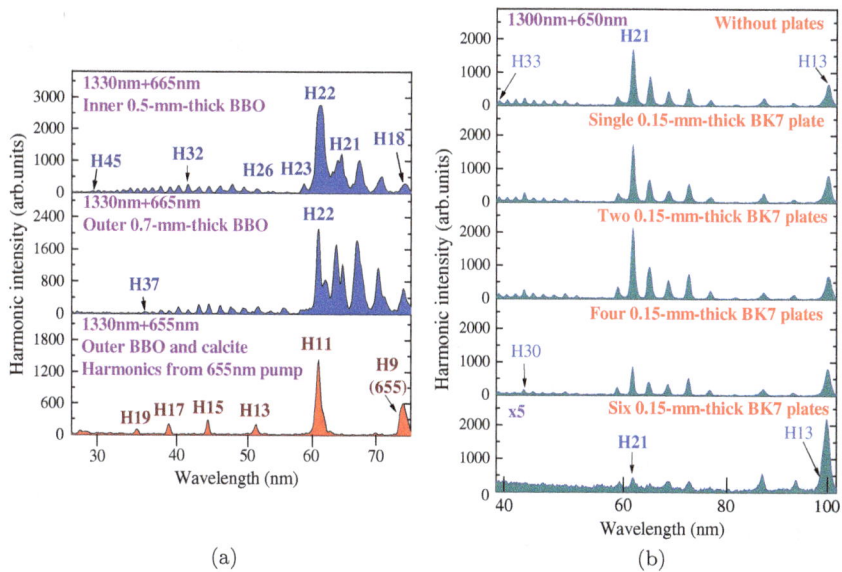

(a)   (b)

Fig. 3.9. (a) Harmonic spectra generated using the BBO crystal inserted inside
(upper panel) and outside (middle panel) the vacuum chamber. Bottom panel
shows generation of the odd harmonics of 655 nm radiation once the calcite plate
was installed in front of BBO crystal, which led to variation of the polarization
of MIR pulse from linear to circular. (b) Harmonic spectra for variable phase
difference between MIR and H2 pulses. One can see a gradual change of the
relative intensities of H13 and resonance-enhanced H21 using different number of
inserted 0.15-mm-thick BK7 plates. Reproduced from [52] with permission from
American Physical Society.

led to generation of high-order harmonics only from the 665 nm pump, which is seen in the bottom graph of this figure, showing the odd harmonics of H2 radiation.

In the two-color pump HHG experiments in gases, the variation of relative phase between fundamental and second harmonic waves of 30 fs pulses allowed observing the 3-fold beatings between the long- and short-trajectory induced harmonics [14]. The use of the focusing optics inside the XUV spectrometer in the experiments did not allow the influence of the short and long trajectories of accelerated electrons on the divergence of harmonics to be distinguished. However, the difference in the plasma harmonic spectra was observed when the thin (0.15 mm) glass plates were introduced between BBO crystal and plasma. The examples of both decrease of the two-color field in the plasma area and decrease of the resonance enhancement of H21 are presented in Fig. 3.9(b). The thin glass plates were inserted after BBO crystal to analyze the variation of the relative phase between two pumps, as in the case described in previous section, and to compare the change of relative intensities of "resonant" (H21) and "nonresonant" (H13) harmonic yields. One can see a significant departure from the large ratio H21/H13 in the case of the absence of the glass plates, which are actually the relative phase modulators [upper panel of Fig. 3.9(b)]), towards the low ratio of these harmonics in the case of propagation through six 0.15-mm-thick plates (bottom panel). Thus the variation of relative phase between pumps may diminish the role of AIS in the single harmonic enhancement. The decrease of pump intensity (due to the growth of Fresnel losses during reflection from a few plates) may also cause a decrease of the overall nonlinear response.

The relation between the addition of a single 0.15-mm thick BK7 plate and relative phase shift between fundamental and second harmonic waves depends on the group velocity dispersion in the glass. The group velocity dispersion leads to a change of the relative phase between waves, and, in the case of thick samples, may lead to the temporal walkoff of two pulses. In particular, due to this effect in the BK7, the 650 nm pulse (2ω) was delayed in the 0.15-mm-thick glass by $\Delta_{BK7} = d[(n_\omega^o)_{group}/c - (n_{2\omega}^e)_{group}/c] \approx 1.3\,fs$ with respect to the 1300 nm pulse (ω) due to $n_{\omega}^{o} < n_{2\omega}^{e}$ in this positive optical element. Here Δ_{BK7} is the delay between two pulses after leaving the thin glass, d is the glass thickness, $c/(n_{\omega}^{o})_{\text{group}}$ and $c/(n_{2\omega}^{e})_{\text{group}}$ are the group velocities of the ω and 2ω waves in the BK7, c is the light velocity, and n_{ω}^{o} and $n_{2\omega}^{e}$ are the refractive indices of BK7 at the wavelengths of the ω and 2ω pumps. In the case of MIR pulses, the group velocity dispersion between fundamental and second-harmonic waves was notably smaller compared with the case of Ti:sapphire laser radiation. Here we also mention the delay between two pulses in the output of BBO crystal. In the case of 0.5-mm-thick BBO, the calculated walkoff between 1300 and 650 nm waves was 24 fs. Note that $t_{2\omega}$ at the output of the 0.5-mm-long BBO crystal was estimated to be 76 fs, while the 1300 nm pulse duration was 70 fs. The additional plates of BK7 led to the change of the relative phase and some decrease of the delay between the envelopes of fundamental and second harmonic waves. The relative phase, ϕ, is proportional to the delay between envelopes, ΔT, and the carrier frequency, $\nu_0(\phi \approx 2\pi\Delta T\nu_0)$ [61]. Thus the single BK7 plate provides $\sim 0.1\pi$ variation of the relative phase.

In the experiments, the HHG occurred in quite a long medium (5 mm). Below we discuss the role of the propagation effect in this kind of experiment. To achieve higher HHG conversion efficiency, the length of the medium might be increased provided that the phase mismatch between the laser field and the harmonic radiation remains low [62]. For the medium lengths where the reabsorption can be neglected and for the optimum phase-matching conditions, the harmonic intensity increases as the square of medium length. However, once the medium length exceeds the coherence length, the harmonic intensity shows oscillations due to phase mismatch [63]. To analyze this propagation process in the LPP, one has to carefully define the best conditions of plasma HHG in the extended medium, while taking into account the peculiar properties of the indium used for laser ablation.

Initially, the dependence of the harmonic yield on the length of indium plasma was analyzed and found that the slope of this curve is close to 2 at the appropriately ablated plasma until maximum length used in these experiments (5 mm). The following studies were carried out under the conditions of this "optimal" LPP. Note that application of stronger ablation of indium target led to decrease of the harmonic yield due to phase mismatch caused by large amount of free electrons. In that case the coherent length of high-order harmonics became shorter than the length of homogeneous extended plasma. The above-mentioned dependence deviated from the slope of 2. To overcome this propagation effect one can use the quasi-phase-matching concept recently demonstrated during HHG studies in the LPP [64,65] based on the formation of the group of short jets instead of the imperforated extended plasma plume.

3.2.3. *Theory of resonance enhancement*

In the following we discuss the results of the numerical simulations of two-color MIR-driven HHG in indium ion. The three-dimensional time-dependent Schrödinger equation (TDSE) was used to describe the interaction between the two-color MIR field and the indium ion (atomic units are used throughout, unless otherwise stated):

$$i\frac{\partial \mathbf{\Psi}(\mathbf{r},t)}{\partial t} = [-\frac{1}{2}\nabla^2 + \mathbf{V}(\mathbf{r}) - \mathbf{E}(\mathbf{r},t)\cdot\mathbf{r}]\mathbf{\Psi}(\mathbf{r},t), \qquad (3.4)$$

where $\mathbf{V}(\mathbf{r})$ is the potential of the plasma system. Here the model potential introduced in [7] was adopted to reproduce the properties of the indium ion:

$$\mathbf{V}(\mathbf{r}) = -\frac{2}{\sqrt{a^2 + \mathbf{r}^2}} + b\exp\left[-\left(\frac{\mathbf{r}-c}{d}\right)^2\right]. \qquad (3.5)$$

This potential can support the metastable state by a potential barrier, which corresponds to the AIS of the ion. The parameters a, b, c and d are chosen to be 0.65, 1.0, 4.0 and 1.6. There is one dominant transition from AIS to ground state ($4d^{10}5s^2\,^1S_0 \rightarrow 4d^95s^25p^1P_1$) in indium ion using this potential [66]. The two-color driving field is synthesized by a x-polarized mid-infrared fundamental field and a y-polarized second harmonic assistant field. The driving field is given by:

$$
\mathbf{E}(\mathbf{r}, t) =
\begin{cases}
E_0 \sin^2(\pi t/T) \cos(\omega_0 t)\mathbf{x} \\
\quad + \sqrt{0.25} E_0 \sin^2(\pi t/T) \cos(2\omega_0 t + \phi)\mathbf{y} & 0 < t < T \\
0 & t > T
\end{cases}
$$

$$(3.6)$$

where E_0, ω_0 are the amplitude and central frequency of the 1300 nm laser field. The carrier-envelope phases of the 1300 nm field and the 650 nm field are set as 0. ϕ is the relative phase between the two fields. T is the pulse duration and is given by $T = 10T_0$, where T_0 is the optical cycle of the 1300 nm pulse. The peak intensities of the 1300 nm pulse and the 650 nm pulse are assumed to be 2×10^{14} and 5×10^{13} W cm^{-2}, respectively. Equation (3.4) can be numerically solved by using the split-operator method [67–69]. Once the evolution of the electron wave function $\Psi(\mathbf{r}, t)$ is found, the time-dependent dipole acceleration $a(t)$ can be calculated with the Ehrenfest theorem [70]

$$
a(t) = -\left\langle \Psi(\mathbf{r}, t) \middle| \frac{\partial[V(\mathbf{r}) - \mathbf{E}(\mathbf{r}, t) \cdot \mathbf{r}]}{\partial r} \middle| \Psi(\mathbf{r}, t) \right\rangle. \qquad (3.7)
$$

The harmonic spectrum $a(\omega)$ is then obtained from the Fourier transform of the dipole acceleration, which is given by

$$
a(\omega) = \int_0^T a(t) \exp(-i\omega t) dt. \qquad (3.8)
$$

The harmonic spectrum with the two-color MIR field is presented by the thick red curve in Fig. 3.10. For comparison, the harmonic spectrum with the fundamental field alone is also presented by the thin blue curve. As shown in Fig. 3.10, the resonant harmonic around 62.2 nm is much more intense than other harmonics. This corresponds to the transition between the ground state and AIS in indium ion. Moreover, adding a second harmonic as the assistant field significantly modifies the spectrum. One can clearly see the enhancement of resonant harmonic yield (the 21st harmonic), the extension of harmonic cutoff, and the generation of even harmonics (the 20th and 22nd harmonics) in the two-color field compared to

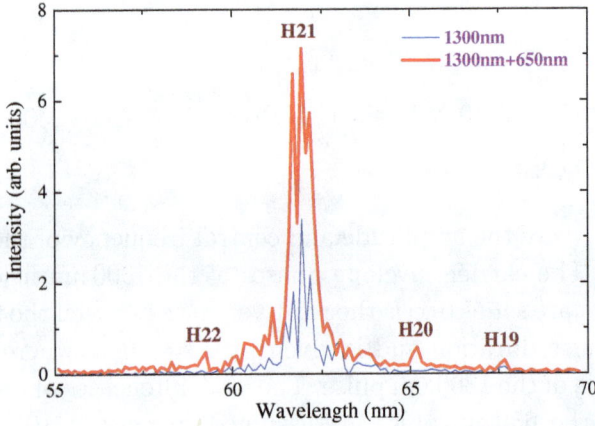

Fig. 3.10. Simulated harmonic spectra from indium ion with single-color (thin blue line) and two-color (thick red line) fields. The relative phase in the two-color field is 0. Reproduced from [52] with permission from American Physical Society.

the single-color case. Figure 3.11 presents the two-color harmonic spectra with different wavelengths of the MIR fields. Other laser parameters are the same with that in Fig. 3.10. It is shown that when tuning the wavelength of the driving pulses, different groups of harmonics around the region of AIS can be selectively enhanced. As the wavelength of the driving pulse varies from 1280 nm to 1380 nm, the tuning of the wavelength of the enhanced radiation is clearly observed, and the resonant harmonic changes from 21st to 22nd order. The theoretical predictions here agree well with the features of two-color resonant HHG observed in experiment as shown in Figs. 3.3(b) and 3.3(c).

In Fig. 3.12(a) we present the variation of intensity of the resonant XUV emission (from $20\omega_0$ to $22\omega_0$) as a function of time in the case of single-color (thin blue curve) and two-color (thick red curve) pumps. Here ω_0 is the frequency of 1300 nm fundamental pulse. One can see that the XUV radiation is delayed with respect to the laser pulse envelope. Almost all of the resonant harmonics are emitted after 5 optical cycles. Besides, there are still XUV emissions after the laser pulse is turned off (from 10 to 20 optical cycles). Such behavior is related to the accumulation of the AIS

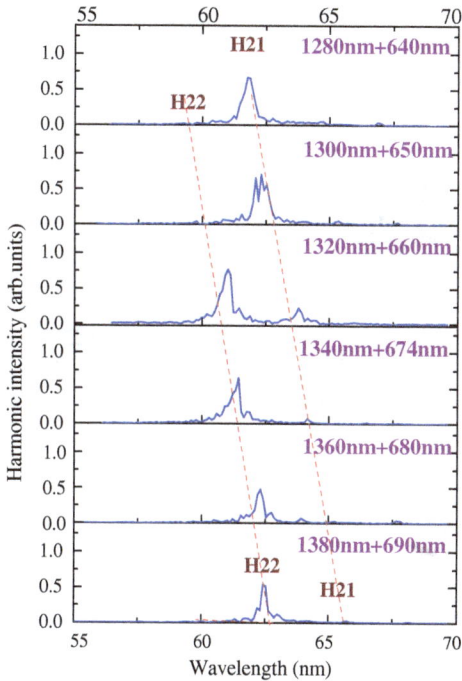

Fig. 3.11. Simulated variations of high-order harmonic yield using different wavelengths of MIR laser fields. Dashed lines show the tuning of the 21st and 22nd orders of harmonics. Reproduced from [52] with permission from American Physical Society.

population in ion system. Compared to the single-color case, the intensity of the resonant harmonic is significantly increased from 12 to 16 optical cycles in the two-color field. In addition, the harmonic yield in the two-color field shows a more stable distribution around the maximum. In the two-color field, the ionization rate of the ground state is increased by the assistant field. Then more ionized and laser-accelerated electrons can be captured into the AIS. Therefore the transition probability between the AIS and ground state is increased and the intensity of the resonant harmonic is enhanced.

It is worth mentioning that the delay time of the two-color resonant harmonics depends on the parameters of two-color field. For example, in Fig. 3.12(a) one can see the XUV emission in the two-color field is more delayed than that in the single-color case with the

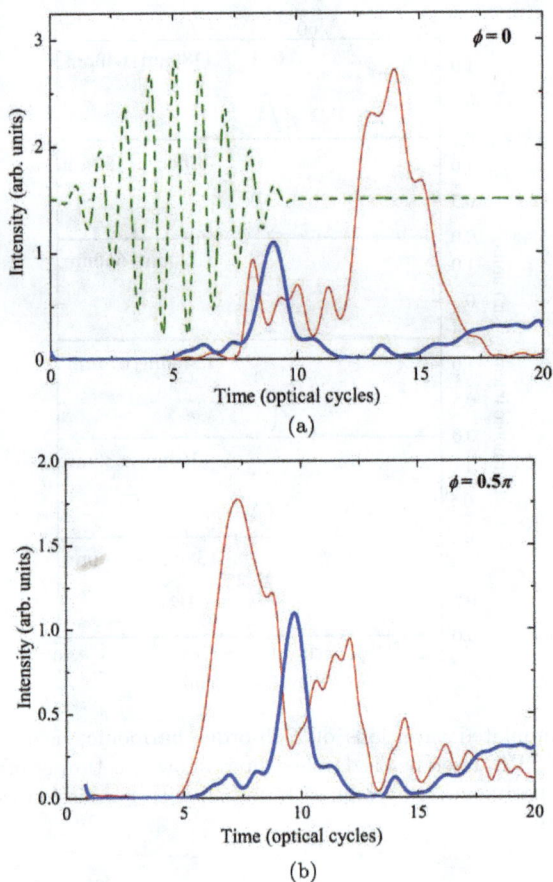

Fig. 3.12. Temporal profiles of resonant emission (from H20 to H22) in the case of single-color (thick solid blue curve) and two-color (thin solid red curve) pumps. The relative phases in Figs. 7(a) and 7(b) are 0 and 0.5π, respectively. The dashed green line shows the electric field of 1300 nm pulse. Reproduced from [52] with permission from American Physical Society.

relative phase $\phi = 0$. But if ϕ varies to 0.5π, the maximum emission of the two-color resonant harmonics occurs at about 7.5 optical cycle, while the maximum emission in the single-color field is at about 9 optical cycle [see Fig. 3.12(b)]. This indicates that the delay effect of the resonant emission in the two-color field is less obvious than that in the single-color field with $\phi = 0.5\pi$. The variation of the temporal profiles of resonant harmonic emission in the two-color field is related to the modulation of autoionization state dynamics by the control field.

The influence of the relative phase on the harmonic yield using the proposed scheme was also investigated. The variations of harmonic yield for the resonant 21st and non-resonant 19th harmonics are presented by the red solid and blue dashed curves in Fig. 3.13(a). The yield has been normalized by the case of $\phi = 0$ for each curve in order to make a clear comparison. Figure 3.13(b) shows the variation of the relative intensities of 21st and 19th harmonics (H21/H19). In Fig. 3.13(c), the comparative harmonic spectra with different relative phases of 0 and 0.5π are presented. One can see that by varying the relative phase of the two-color field one can influence the yield of the resonant and non-resonant harmonics. It was found that the change of the harmonic yield in the resonance region is more significant than that in the non-resonance region. When ϕ varies from 0 to 0.5π, the yield of the 21st harmonic decreases faster than that of the 19th harmonic. Correspondingly the ratio of relative intensities for H21/H19 changes from large values to smaller ones. This indicates that the relative phase of the two-color field may play a role in single harmonic enhancement during resonance-affected HHG. Besides, it was found that the variation of the harmonic spectrum at different ϕ is related to the pulse envelope used in the calculations. When using the trapezoid envelope, the influence of ϕ on the harmonic spectrum is smaller than that using 'sin^2' envelope.

3.2.4. *Discussion*

The main message of this work is the first demonstration of the use of tunable sources for analysis of the influence of strong resonance on the harmonic efficiency for different orders, contrary to previous applications of the sources with fixed wavelengths, which do not allow the nonlinear spectroscopy of ionic media. The effect of the resonance on the HHG, though not new, is still debated in literature (see for example [71]). Note that no resonance enhancement of harmonics was reported using gaseous media, while harmonic generation in the plasmas is a proven road for the analysis of the spectroscopic features

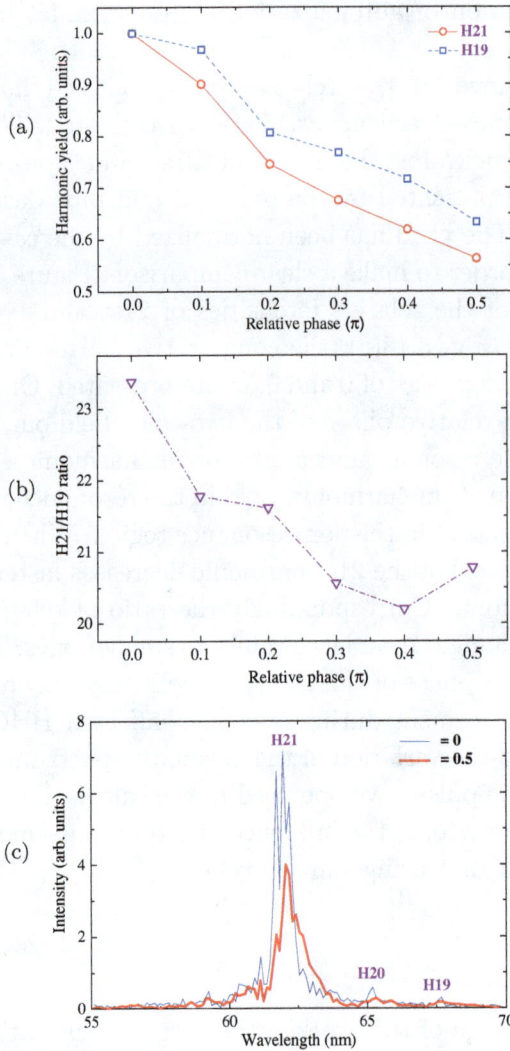

Fig. 3.13. (a) Dependences of H19 and H21 yields on the relative phase between the 1300 nm and its second harmonic pulses. (b) Variation of the relative intensities of the 21st and 19th harmonic with relative phase. (c) Simulated harmonic spectra with different relative phases of 0 and 0.5π. Reproduced from [52] with permission from American Physical Society.

of various ionic species using the method of high-order nonlinear spectroscopy. Needless to say this method allows high fluencies of XUV photons to be achieved. In [24], the difference with previous studies of the influence of resonances on the HHG was shown. In fact, there are no such studies which analyze this influence, since no tunable laser sources were applied for the HHG in the plasmas, excluding [72]. Meanwhile, the application of multi-cycle pulses from OPA for gas harmonics has shown the perspectives for generation of attosecond pulses. Thus, the proposed approach could be considered as a road for overcoming the restriction in generation of ultrashort pulses. Plasma harmonics, as an alternative to gas ones, has already showed many advantages which the latter method does not have. The use of OPA allowed, as has been shown in our manuscript, further steps in understanding of the peculiarities of resonance enhancement to be achieved.

The application of two-color pump allowed a significant enhancement of harmonic yield compared with single-color pump. This claim is not a new one, since it has been proven by many researchers. The novelty is the use of this feature for the analysis of resonant HHG in the mid-infrared field, which cannot be realized using a single-color scheme due to the λ^{-5} rule. The use of only fundamental pump is almost impossible and unpractical for the tasks of this research, since no strong harmonics were observed at all, or they were extremely weak. The use of second wave is not a trick but rather a necessity to study the resonant high-order harmonic generation using the mid-infrared pump.

Moreover, the changes in the properties of these two pumps were made, which allowed observation of the variable response of the medium. One of these changes was the modification of the relative phase between two pumps. This modification resulted in the significant change in the relative intensities of resonance-enhanced harmonic and other ordinary harmonics (compare upper and bottom panels of Fig. 3.9b). This finding clearly indicates the stronger influence of the relative phase between pumps on the resonance-enhanced harmonics rather than "non-resonant" harmonics. The calculations of the role of relative phase showed that it was less strong than the experiment shows. The importance of these data is underlined by the fact that they show the relative influence of the phase on the "resonance" and "non-resonant" harmonics, since the additional impeding factors (optical losses, imperfect overlap of pulses, etc.) equally influence both harmonics.

The variation of the temporal profiles of resonant emission in the two-color field is related to the modulation of resonant HHG process by the control field. However, it is very difficult to fully analyze the behavior of autoionizing state in the exterior laser field (variation of energy, width, decay time, etc.), because this information is entangled in the evolution of the wave function in our current TDSE model. The proposed theoretical model (i.e., numerical solving of the 3D TDSE) allowed us to reproduce the experimental observations and characterize the new spectral and temporal characteristics of the two-color resonant HHG compared to the single-color case. These results indicate that the dynamics of the autoionization state can be turned by the two-color field, therefore the temporal profile of the resonant emission in the two-color field depends on the relative phase.

Since the role of micro- and macro-processes in resonance enhancement is still debated, this result points out the stronger influence of the former processes. The interpretation of these results probably requires additional studies. Notice that the study [71] also claims the same, while analyzing the variations of the coherence length of these harmonics. Thus the analysis of resonance processes by different means may offer different options in definition of the relative influence of single particle response and collective response of the medium on the harmonic yield.

The HHG efficiency in the indium LPP was optimized with respect to various parameters. In particular, small delay between 350 ps heating pulses and 70 fs driving pulses (<5 ns) did not allow the observation of harmonics at the used distance between the target and the axis of driving beam propagation. The harmonic generation efficiency abruptly increased once the delay exceeded 5 ns. In the case of indium plasma, the maximal harmonic yield was observed at 35 ns delay. At longer delays, for a fixed distance between the target and the laser beam, the harmonic yield started gradually decreasing until it disappeared entirely at $\sim$150 ns. However, the conversion efficiency was able to be optimized for the delays larger than 35 ns by increasing the distance between the target and the laser beam. In the discussed studies, the maximum efficiency of HHG for a 35-ns delay was achieved when the distance between the focal region of the driving beam and the target surface was $\sim$150 μm.

The optimal delay depends on the target material, in particular on its atomic number (Z). Different targets were analyzed to reveal the optimal plasma medium at a certain delay between the pulses, corresponding to the maximum conversion of femtosecond radiation into the harmonics. It was shown that, at relatively short delays, targets with smaller Z values provide higher conversion efficiency in comparison with heavy targets due to larger velocities of the former particles. These measurements were performed at 20-ns delay between the ablated and converted pulses. One might expect this value to be optimal for light ions and atoms because, for heavier ions such as indium (Z $=$ 49, atomic weight 115), the time of flight from the target surface to the region of femtosecond radiation propagation exceeds the delay between the pulses because of the lower velocities of In ions in comparison with lighter species.

Below, we briefly discuss the plasma formation above the indium target surface. This process cannot be explained by simple heating of the target surface, its successive melting and evaporation, and the runaway of particles with thermodynamic velocities. These relations are valid for the fairly slow processes induced by long pulses, where the atomic velocity at a target heating temperature of 1000 K is about 7×10^2 m s^{-1}. Under these conditions, atoms and ions move by only 15 μm from the target surface during 20 ns. On the assumption that laser plasma formation is determined in our case by such a slow process, the generation of harmonics would be impossible because the radiation to be converted passes at a distance of 150 μm above the target surface. In this case, the particles would reach the interaction region only 200 ns later. At the same time, efficient generation of harmonics (e.g., in indium plasma) started to be observable even at a delay of 20 ns between the pulses and barely occurs at delays above 150 ns. This also holds true for other plasmas.

This inconsistency between the thermal model of propagation of evaporated material and the observed efficient HHG in LPP at small delays between the ablating and driving pulses is explained within another model of plasma formation, specifically, the plasma explosion during target ablation by short pulses. The dynamics of plasma front propagation during laser ablation by short pulses has been analyzed in a number of studies (see, e.g., [73, 74] and references therein). A numerical analysis of plasma formation from a target irradiated with a single laser pulse was reported in [73]. The velocities of plasma front in accordance with plasma explosion model were in the range of 1×10^4 to $1 \times 10^5 \, \mathrm{m \, s^{-1}}$.

In experiment, the dynamics of plasma origin and propagation can be analyzed using the shadowgraph technique. The spatial characteristics of the laser plasma formed by short pulses under similar conditions on the surfaces of lighter (boron, $Z = 5$) and heavier (manganese, $Z = 25$) targets were reported in [75]. In the case of heavier target (Mn), the plasma front propagated at a velocity of $6 \times 10^4 \, \mathrm{m \, s^{-1}}$. The plasma front passed a distance of 130 μm after a few nanoseconds, rather than several hundreds of nanoseconds, in accordance with the above-described thermal model of plasma expansion. One can assume that, in the case of In ions possessing twice the mass of Mn, the slower movement causes later appearance of the plasma cloud in the area of laser beam propagation. Obviously, the formation of optimal plasma is not limited by the occurrence of plasma front in the propagation region of radiation to be converted. To this end, some time is necessary for the concentration of particles responsible for harmonic generation to reach a certain value.

The application of fundamental waves together with assistant field dramatically changes the efficiency of HHG and allows the observation of the resonantly enhanced harmonic alongside with the enhanced nearby harmonics. The influence of assistant (H2) field on HHG did not crucially depend on the relative intensities of two pumps. In the discussed studies, the maximal efficiency of second-harmonic generation using the 1310 nm pulses (27%) means the ratio of the second-harmonic pulse energy and a whole 1310 nm pulse energy (measured before the propagation of BBO).

This efficiency shows that the ratio of assistant (655 nm) and driving (1310 nm) pulse energies inside the plasma was approximately 1:3. The theoretical calculations took into account different ratios of two interacting pulses. The variation of this ratio between 1:5 and 1:3 did not definitively change the conclusions defined from the calculations of the used theoretical model. In present studies, we showed the calculations of harmonic spectra using the 1:4 ratio. The role of second field in the used ratio of pulse energies was similar to the influence of this assistant field at rather smaller ratios. In particular, in the case when the 1:5 and 1:20 ratios were used, similar strong enhancement of the odd harmonics, as well as equal odd and even harmonic intensities were achieved in previous studies using the 800-nm-class lasers [76, 77].

As already mentioned, the first observation of resonant HHG was reported in [1] for the low-charged indium plasma prepared by laser ablation. In those and other experiments with indium plasma [66, 78–80], the thirteenth harmonic of 800 nm laser radiation was a few tens of times more intense than neighboring harmonics. This phenomenon is attributed to a strong multiphoton resonance with exceptionally strong transition of single-charged indium ion, which can easily be Stark-shifted toward the thirteenth harmonic of 800 nm pump radiation.

The enhancement of the 13th harmonic emission from the indium plasma observed in those and present studies is due to the influence of radiative transitions between the $4d^{10}5s^2\,{}^1S_0$ ground state of In II and the low lying $4d^95s^2\,np$ transition array of In II. Among them, the transition at 19.92 eV (62.24 nm) corresponding to the $4d^{10}5s^2\,{}^1S_0 \rightarrow 4d^95s^25p\,{}^1P_1$ transition of In II is exceptionally strong. The gf value of this transition, the product of the oscillator strength f of a transition and the statistical weight g of the lower level, has been calculated to be $gf = 1.11$ [81], which is more than twelve times larger than that of any other transition from the ground state of In II. This transition is energetically close to the 13th harmonic ($h\nu_{13H} = 20.15$ eV or $\lambda = 61.53$ nm) of 800 nm radiation and the 21st harmonic ($h\nu_{21H} = 20.02$ eV or $\lambda = 61.9$ nm) of 1300 nm radiation, thereby resonantly enhancing its intensity.

3.3. Resonance Enhancement of Harmonics in Laser-produced Zn II and Zn III Containing Plasmas

3.3.1. *Method of laser ablation induced MIR-pumped HHG spectroscopy*

Studying materials using the laser ablation-induced high-order harmonic generation spectroscopy, which exploits the spectral properties of various solids through their ablation and further propagation of the short laser pulses of various wavelengths through the laser-produced plasma leading to generation of high-order harmonics in the extreme ultraviolet range, is an attractive way for application of the nonlinear optical processes for the definition of the material properties. To distinguish one material from another one has to use the specific properties of matter, particularly their map of the energetic levels of excited ionic states. This approach can be accomplished through the resonance enhancement of some high-order harmonics in the vicinity of ionic transitions, especially those possessing large oscillator strengths.

Resonance enhancement of single high-order harmonics in various laser-produced plasmas has been reported since the first observation of this process in indium plasma [1]. Tellurium, chromium, manganese, arsenic, selenium, tin, and antimony were among the plasma media where the yield of some harmonics was higher than one from the lower-order harmonics, contrary to the commonly accepted rule of a decrease of conversion efficiency for each next order of harmonics. Those studies were carried out using fixed wavelengths of the driving pulses of Ti:sapphire lasers. The lack of tunability did not allow the analysis of the nonlinear optical processes at the conditions of fine tuning of driving pulses for the growth of single harmonic conversion efficiency.

The change of excitation of matter may cause the appearance of some additional excited states, which can enhance the neighboring harmonics as well, which makes it difficult to analyze the complex behavior of harmonic generation in LPP. Among such species, the zinc plasma can be considered as an interesting subject of studies, since it allows the conditions of excitation of AIS to be varied depending on the conditions of ablation of the bulk target. The variation of target excitation allows the analysis of the nonlinear response of such plasma by comparing the enhancement of different harmonics provided they coincide or stay close to those resonances.

One can conclude from above consideration that in order to facilitate the use of plasma harmonic concept for laser-ablation induced HHG spectroscopy one should use tunable sources of laser radiation allowing the fine tuning of driving pulses and correspondingly harmonic wavelengths along the spectral ranges of strong ionic transitions. In this section, we analyze the fine tuning of harmonics in the vicinity of neutral and ionic transitions of zinc during HHG using mid-infrared source of ultrashort pulses and its second harmonic propagating through the LPP. We analyze the enhancement of those harmonics depending on the plasma formation conditions [82].

3.3.2. *Experimental conditions of HHG in zinc plasma using tunable MIR pulses*

The experimental setup consisted of Ti:sapphire laser, travelling-wave optical parametric amplifier (OPA) of white-light continuum, and high-order harmonic generation scheme using propagation of amplified signal pulse from OPA through the extended LPP (see subsection 3.1.2 for details). The mode-locked Ti:sapphire laser pumped by diode-pumped, cw laser was used as the source of 803 nm, 55 fs, 82 MHz, 450 mW pulses for injection in the pulsed Ti:sapphire regenerative amplifier with pulse stretcher and additional double passed linear amplifier. The output characteristics from this laser were as follows: central wavelength 806 nm, pulse duration 350 ps, pulse energy 5 mJ, 10 Hz pulse repetition rate. This radiation was further amplified in home-made Ti:sapphire linear amplifier up to 22 mJ. Part of this radiation with pulse energy of 6 mJ was separated from a whole beam and used as a heating pulse for homogeneous extended plasma formation using the 200-mm focal length cylindrical focusing lens installed in front of the extended solid target placed in the vacuum chamber [Fig. 3.14(a)]. The intensity of the heating pulse

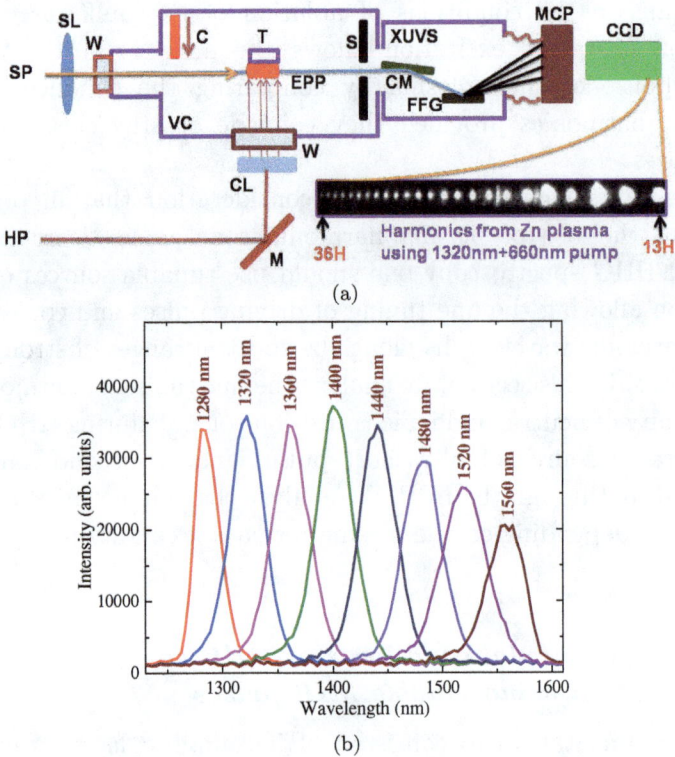

Fig. 3.14. (a) Experimental scheme for harmonic generation in zinc plasma using tunable MIR radiation. SP, driving signal pulse; HP, picosecond heating pulse; SL, spherical lens; CL, cylindrical lens; M, mirror; VC, vacuum chamber; W, windows of vacuum chamber; T, zinc target; C, nonlinear crystal; EPP, extended plasma plume; S, slit; XUVS, extreme ultraviolet spectrometer; CM, gold cylindrical mirror; FFG, flat field grating; MCP, micro-channel plate; CCD, charge coupled device camera. Inset: raw image of harmonic spectrum from zinc ablation using the two-color MIR pulses (1320 nm and 660 nm). (b) Tuning spectra from optical parametric amplifier at different central wavelengths (marked on the top of each spectrum). Reproduced from [82] with permission from IOP Publishing.

on the target surface was varied up to $4 \times 10^9 \, \mathrm{W \, cm^{-2}}$. The ablation sizes were $5 \times 0.08 \, \mathrm{mm^2}$.

The remaining part of amplified radiation was delayed with regard to the heating pulse in such a way that, after compression and pump of OPA, the signal pulse from parametric amplifier propagated through the formed plasma plume 35 ns from the beginning of target ablation. After propagation of compressor stage the output characteristics of a whole Ti:sapphire system were as follows: pulse energy 8 mJ, pulse duration 64 fs, 10 Hz pulse repetition rate, pulse bandwidth 17 nm, central wavelength 806 nm. This radiation pumped the OPA HE-TOPAS Prime (Light Conversion). Signal and idler pulses from OPA allowed tuning along the 1200–1600 nm and 1600–2600 nm ranges respectively (Fig. 3.14a).

In the HHG experiments we used the signal pulses, which were 1.5 times stronger than the idler pulses. Most of experiments were carried out using the 1-mJ, 65-fs signal pulses tunable in the range of 1250–1400 nm, which was sufficient for the tuning along the resonances. Spectral bandwidth of tunable pulses was 45 nm [Fig. 3.14(b)]. The intensity of the 1310 nm pulses focused by 400 mm focal length lens inside the extended plasma was 2×10^{14} W cm^{-2}. The driving pulses were focused onto the extended plasma from the orthogonal direction, at a distance of $\sim 100 \mu$m above the target surface [Fig. 3.14(b)]. Most of experiments were carried out using the two-color pump of LPP. The plasma and harmonic emissions were analyzed using an XUV spectrometer containing a cylindrical mirror and a 1200 grooves/mm flat field grating with variable line spacing. The spectrum was recorded on a micro-channel plate detector with the phosphor screen, which was imaged onto a CCD camera.

The Zn plasma was analyzed as the medium for harmonic generation. Previously, this plasma has shown some attractive features, particularly the single harmonic enhancement using a few-cycle pulses [83] and a high low-order harmonic yield in the case of multi-cycle pulses [84]. In both cases the fixed wavelength of Ti:sapphire laser was used for the excitation of harmonics in the Zn plasma. The 5-mm-long zinc plate was installed in the vacuum chamber for laser ablation.

3.3.3. *Single- and two-color pumps of zinc plasma*

The harmonic yield using MIR pulses (1 mJ, 1300 nm) was significantly weaker compared with the 8 mJ, 806 nm pump due to the

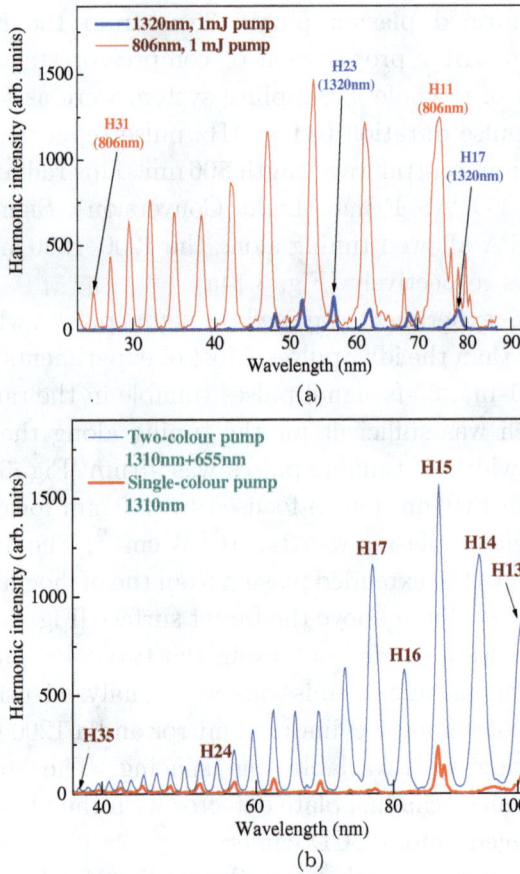

Fig. 3.15. (a) Harmonic spectra from zinc plasma using 806 nm (thin red curve) and 1320 nm (thick blue curve) pulses measured under similar conditions. (b) Comparative spectra from Zn plasma in the case of single-color (thick red curve) and two-color (thin blue curve) pumps under similar conditions of experiment. Reproduced from [82] with permission from IOP Publishing.

$I_H \propto \lambda^{-5}$ rule. The comparison of these two pumps at similar energies of pulses (1 mJ) showed an eight-fold decrease of harmonic yield from zinc plasma in the case of 1320 nm pulses compared with the 806 nm pulses [Fig. 3.15(a)], close to the theoretical prediction of this ratio (11.8). The harmonic cut-off in the case of MIR pulses was also lower compared with the 806 nm pulses, contrary to the expectations in the extension of cut-off for the longer-wavelength pump ($E_{\text{cut-off}} \propto \lambda^2$), due to very small conversion efficiency, which did not allow the observation of harmonics below the 40 nm spectral region.

Thus, the use of 1 mJ MIR pulses for plasma harmonic studies caused generation of weak odd harmonics in the longer wavelength range of extreme ultraviolet. Because of this the two-color pump of plasma was used. This scheme allowed the odd harmonic yield to be enhanced in the case of 800 nm + H2 pump of plasmas [77, 85]. In those studies, an 8-fold growth of harmonic yield was reported compared with the single-color (800 nm) pump of plasmas.

In present studies, the insertion of the BBO crystal on the path of focused driving beam drastically modified the harmonic spectrum. The extension of harmonic cut-off compared with single-color pump, 6× to 10× growth of the yield of odd harmonics, comparable harmonic intensities for the odd and even orders along the whole range of generation, and tuning of harmonics allowing the optimization of resonance-induced single harmonic generation were among the advanced features of these experiments [Fig. 3.15(b)]. The most important feature of these experiments was the availability in the analysis of the influence of excited states on the harmonic distribution under different excitation conditions of ablating target. Below, we present the results of harmonic spectra modification using tunable driving two-color (MIR + H2) orthogonally polarized pulses leading to the observation of resonance enhancement of some harmonic orders caused by different species of plasma.

The advantage of the studies of resonance-induced enhancement of harmonics using tunable MIR + H2 pulses is the opportunity of fine tuning of this high-order nonlinear optical process for spectral enhancement of the harmonic yield. Another advantage is closely related with the analysis of the oscillator strengths of some ionic transitions of metals and semiconductors using the HHG approach. Thus this approach could be considered as a new method of nonlinear spectroscopy. Here we show some examples of resonance enhancement of the harmonics of MIR radiation using the Zn target ablated at different heating pulse fluencies. The use of tunable broadband MIR radiation allowed the analysis of the enhancement of the groups of harmonics close to the resonances possessing strong oscillator strengths.

3.3.4. *Modification of harmonic spectra at excitation of neutrals and doubly charged ions of Zn*

Different fluencies of heating pulses on the Zn target allowed formation of the LPP containing various species (i.e., excited neutrals, singly and doubly charged ions, and electrons). The aim of reviewed studies was to analyze the dynamics of harmonic spectra at weak, medium, and strong ablation of zinc by 350 ps pulses. These terms mean the conditions of plasma, when harmonics are not cancelled by a large amount of free electrons leading to phase mismatch but rather influenced by the presence of different species in the plasma. We controlled plasma excitation by analyzing the emission spectra in the visible and XUV ranges. At weak ablation of target (i.e., at a fluency of $0.6\,\mathrm{J\,cm^{-2}}$), no XUV emission lines were seen in those spectra. Under these conditions, a set of harmonic spectra was measured using the tuning of MIR pulses with a step of 30 nm. The HHG spectra demonstrated the featureless pattern when each next (odd and even) order gradually decreased starting from longest observable harmonic (H13) up to H24 [Fig. 3.16(a)], while tuning the MIR pulses between 1280 and 1460 nm. Thus weak excitation of Zn did not lead to the formation of the conditions for the resonance-induced enhancement of single harmonic.

The growth of fluency ($0.8\,\mathrm{J\,cm^{-2}}$) led to appearance of the emission of excited singly charged zinc ions. This plasma demonstrated emission of the group of Zn II lines at the wavelengths of 75.5, 76.7, and 77.9 nm, as well as 88.1 and 89.3 nm. Under these conditions, the plasma harmonic spectra started to modify and show the enhancement of the harmonics tuned along the 75.5–77.9 nm range [Fig. 3.16(b)]. The third panel from the top of this figure shows the enhanced H17 compared with H16 and H18. The enhancement factor of this harmonic varied between 1× and 8× depending on the wavelength of MIR pulses, while other harmonics showed the featureless plateau-like spectrum. The variation of the wavelength of MIR

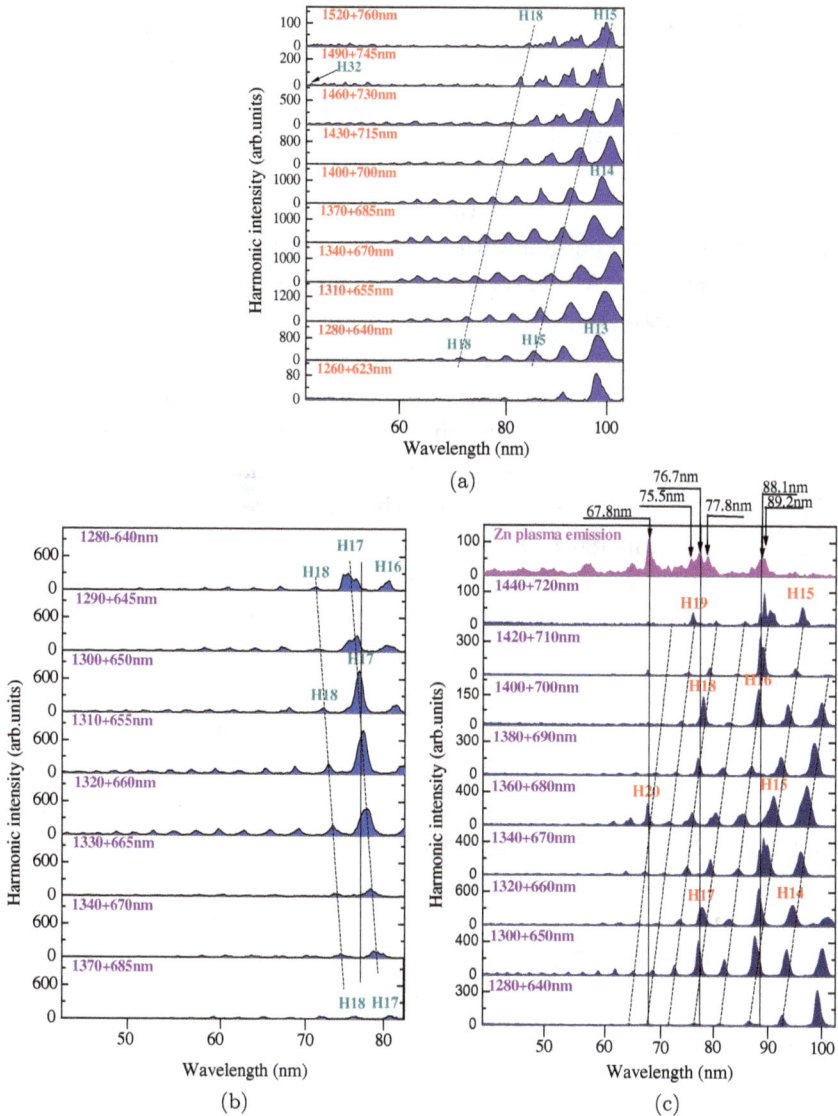

Fig. 3.16. Tunable harmonic spectra in the cases of (a) weak, (b) moderate, and (c) strong excitation of Zn target (see text for further details). dashed lines show the tuning of harmonics. Solid lines show the ionic transitions responsible for enhancement of harmonics. Reproduced from [82] with permission from IOP Publishing.

pulses with the step of 10 nm allowed the variation of the wavelength of H17 with the step of 0.6 nm. With this tuning, the maximally enhanced harmonic occurred when its wavelength coincided with the strongest emission line (third panel from the top, 76.7 nm [86]).

Further growth of fluency $(1.3\,\mathrm{J\,cm^{-2}})$ led to appearance of strong emission lines [Fig. 3.16(c), upper panel]. Under these conditions, the harmonic generation deteriorated due to large amount of free electrons. A decrease of excitation of target (down to $\sim1\,\mathrm{J\,cm^{-2}}$) allowed generation of harmonics in the presence of plasma emission lines. The observed HHG spectra under these conditions showed the resonance-induced enhancement at different wavelengths corresponding to different transitions. In particular, the 1300 nm + 650 nm pump of plasma led to generation of enhanced H15 and H17. The enhancement originated from the influence of Z II resonances (76.7 and 88.1 nm [86]). Moreover, the appearance of a strong line of Zn III [67.8 nm [87], see upper panel of Fig. 3.16(c)] led to enhancement of harmonics in the vicinity of this spectral region. In particular, it was clearly seen in the case of 1360 nm + 680 nm pump, when H20 ($\lambda = 68$ nm) almost coincided with the above ionic line. One can see the 5-fold enhancement of H20 with regard to the neighboring harmonics. At the same time, this spectrum shows the absence of enhancement of the harmonics near Zn II lines (76.5 and 88.9 nm), since they were detuned from these regions. Note that the enhancement of the part of broadband H11 of a few-cycle (4.5 fs) pulses in the vicinity of 67.8 nm (18.3 eV, $3\mathrm{d}^{10} \rightarrow 3\mathrm{d}^9(^2\mathrm{D})4\mathrm{p}$) was reported previously in [83].

These experiments have demonstrated that the resonance enhancement of nearby harmonic close to the resonance can be observed in the case of sufficient amount of excited species (i.e., singly or doubly changed ions of zinc in our case). Thus the availability of resonance enhancement depends on the population of appropriate energetic levels of ions. Note that only part of the harmonic was enhanced in the vicinity of 67.8 nm. Figure 3.17 shows the emission spectrum of H18–H22 in the case of 1350 nm + 675 nm pump. One can see that the bandwidth of enhanced 20th harmonic (~0.4 nm) was ~4 times narrower than the one of the neighboring harmonics

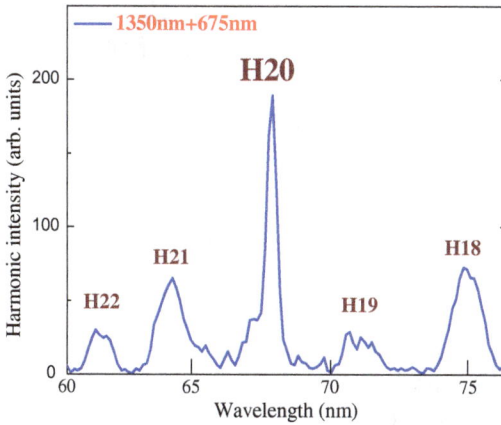

Fig. 3.17. Enhancement of the part of H20 under the conditions of coincidence of this harmonic with the 67.8 nm transition of Zn III. Reproduced from [82] with permission from IOP Publishing.

($\sim$1.6 nm). This enhancement of the part of harmonic shows a similarity with the recently reported partial enhancement of single harmonic in Ar gas, which was attributed to the influence of Fano resonance [31]. There are also a few other assumptions, which will be discussed in the following subsection.

3.3.5. *Discussion of results*

Recent studies have shown the advanced properties of extended Zn plasma for generation of the strongest lower-order harmonics of Ti:sapphire laser among other plasma species [83]. The range of energy fluencies of $\sim$0.3–1 J cm^{-2} seems to satisfy the requirement for the formation of the optimal plasma, which has previously allowed the generation of highest harmonic orders. Those studies have also shown a strong emission of 9th harmonic (89.1 nm) of Ti:sapphire laser, which considerably exceeded the neighboring orders. The ratio of the intensities of this and neighboring higher-order harmonics ($\sim$8$\times$) analyzed at unsaturated conditions of registration was considerably larger compared with other plasmas (C, Au, Cu). This observation points out that the mechanism of the enhancement of 89.1 nm radiation was other than the prevalence of the lower orders over the higher orders at the beginning of the plateau-like range of harmonic distribution. Analysis of plasma emission during excitation of the Zn target, without the propagation of the driving pulse through the plasma, showed the presence of some ionic lines attributed to the Zn II and Zn III transitions. Note the observed closeness of two $3d^{10}4s-3d^94s4p$ transitions of Zn II (88.1 and 89.3 nm) with the wavelength of 9th harmonic of 802 nm driving radiation.

Further, the recently reported observation of strong emission of the 18.3 eV transition $(3d^{10}-3d^9(^2D)4p)$ of Zn III under the conditions of plasma excitation by a few-cycle broadband pulses centered at 770 nm [83] was attributed to the enhancement of the part of 11th harmonic of this radiation, though the wavelength of this harmonic (70 nm) did not exactly match with the wavelength of $3d^{10}$ the $3d^9(^2D)4p$ transition (67.8 nm). In those studies, a narrowband enhanced emission was similar to the one observed in the reviewed experiments [82] and could also be attributed to the influence of resonances and propagation processes. As we already mentioned, there are some other explanations of the resonance enhancement of harmonics, which are mostly based on the analysis of the micro- and macro-processes. Below we address various options for explanation of resonance enhancement of harmonics.

The autoionizing levels of Zn have been reported in a few studies [86–92]. Figure 3.18 summarizes published data on the lines in Zn atoms, singly and doubly charged Zn ions in the 50–100 nm spectral region. To make the comparison more obvious the spectral regions where the lines are present or absent are shown, but not the positions of all the numerous lines. One can see that the HHG enhancement in the spectral region 67.6–72.5 nm can take place only in Zn III. So the enhanced coherent XUV pulses at 67.8 nm in Fig. 3.16(c) show the presence of essential ratio of these ions in plasma under stronger excitation. The absence of the enhancement at this wavelength in Figs. 3.16(a,b) shows that the number of such ions is negligible under weak and moderate plasma excitation. The enhancement in the spectral region longer than 74.3 nm can be due to Zn II ions but not Zn III; this enhancement is observed in

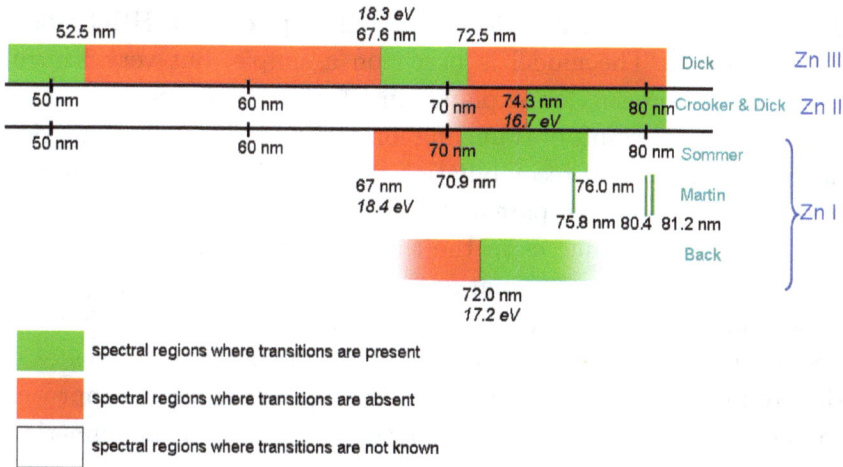

Fig. 3.18. Spectral regions of lines corresponding to the transitions to excited states (in particular, to AIS) in Zn I, Zn II, Zn III according to the published data (from top to bottom [86–88, 90, 91] respectively). Reproduced from [82] with permission from IOP Publishing.

both cases of moderate and strong excitation, showing that in both cases an essential number of Zn II ions is present in the plasma. The disappearance of the resonant enhancement for the case of weak excitation in Fig. 3.16(a) shows that (i) Zn II ions are almost absent in this case and (ii) the Zn neutrals (which thus dominate in the plume) do not provide resonant HHG enhancement. Note that Zn has relatively high first ionization potential (9.4 eV), so domination of neutrals in the laser plume under weak excitation is very natural; this is not the case for other metals used in plasma HHG experiments.

In Fig. 3.18, we show that transitions to excited states (in particular, to AIS) corresponding to wavelengths longer than 71–72 nm are present in neutral Zn. However, these transitions do not provide HHG enhancement in Fig. 3.16(a). This can be attributed to the low oscillator strengths of these transitions for wavelengths shorter than 104 nm [93].

There are a number of theoretical approaches describing the influence of resonances at the microscopic response [5–7, 9, 10, 47,

48, 50, 51]. In particular in Ref. [7], a four-step resonant HHG model was suggested. The model is based on a simple, but very fruitful three-step approach for explanation of the HHG [38, 39], which describes it as a result of tunneling ionization, free-electronic motion in the laser field, and recombination accompanied by XUV emission upon the return to the parent ion. In the four-step model, the first two steps are the same as in the three-step one, but instead of the last step (radiative recombination from the continuum to the ground state) the free electron is trapped by the parent ion, so that the system (parent ion + electron) lands in the AIS, and then it relaxes to the ground state emitting XUV. In [7, 9, 10], the numerical approach to study resonant HHG was developed. It is based on the numerical solution of the TDSE for an electron in a model potential and the external laser field. The model potential has an excited quasi-stable state, which models the AIS in the real atom or ion.

In this section, we analyzed the approach developed to study numerically the HHG in Zn ions. Note, that in the reviewed study the resonances were observed, which were due to transitions to the highly excited bound states, not to the AIS. However, these bound states are broadened due to photoionization in the intense laser field. So in this sense there is no fundamental difference between transitions to AIS (enhancing above ionization threshold harmonics) and transitions to highly excited bound states (enhancing below-threshold harmonics). Moreover, the generation of the below-threshold harmonics also includes the quasi-free electronic motion as one of the steps of the process [94, 95], so the resonant enhancement of this generation can be described within the similar four-step model.

The model potential suggested in [7] was used to reproduce the resonant HHG enhancement by Zn II ions near 76.7 nm. The TDSE is solved numerically for the electron in this potential and the external laser field with intensity of $2 \times 10^{14}\,\mathrm{W\,cm^{-2}}$ and different wavelengths. The calculated spectra are shown in Fig. 3.19. One can see that the enhancement is maximal when the harmonic frequency coincides with the transition one, and substantially decreases when the fundamental wavelength is detuned by only 10 nm. So the calculated results reproduce reasonably well the experiment presented

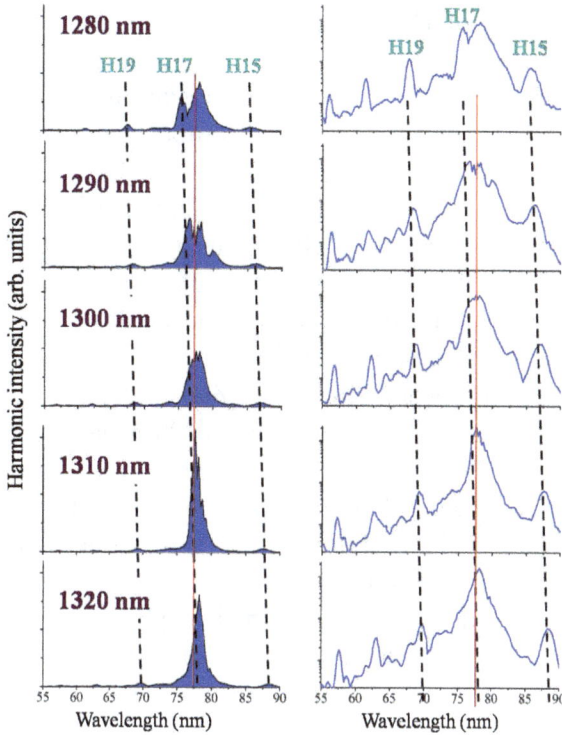

Fig. 3.19. The harmonic spectra calculated via TDSE numerical solution in the linear (left) and logarithmic (right) scales. The fundamental wavelengths are presented in the graphs. The dashed lines show the tuning of harmonics, the solid line shows the ionic transition. Reproduced from [82] with permission from IOP Publishing.

in Fig. 3.16(b). The calculated enhancement is higher than the observed one.

The width of the resonantly-enhanced H17 is close to the ones of the other harmonics, both in calculated and experimental result. This is not the case for the experimentally observed enhancement of H20 in Zn III, see Fig. 3.17. This difference can be explained using an analytical theory based on the four-step model [10]. In particular, this theory shows that the shape of the resonantly enhanced harmonic line is a product of the harmonic line, which would be emitted in the absence of the resonance, and the enhancement factor, which essentially exceeds 1× within the spectral width of the excited state.

The enhancement of H17 in Zn II is due to this state, which is very close to the ionization threshold (the difference is approximately 2 eV), so it was broadened due to photoionization. Thus the enhancement takes place in the spectral region comparable with the harmonic width, and the latter remains approximately the same as for the non-resonant harmonics. The state in Zn III enhancing H20 is 20 eV lower than the ionization threshold, so it remains narrow. The enhancement region is narrower than the harmonic line width, so the H20 line shape is close to one of the surrounding harmonics but with a narrow maximum due to the resonance enhancement.

The narrow plasma emission line of Zn II (67.8 nm) coincides with the central wavelength of the broadband 20th harmonic originated from 1350 nm + 675 nm two-color pump. One can assume that the part of this harmonic, being in resonance with the above ion transition, became enhanced due to resonance-induced increase of the nonlinear optical response of the plasma. In that case, propagation effects may play a decisive role for efficient phase matching between the driving and harmonic fields in the vicinity of the above-mentioned ionic transition.

Among the factors responsible for the enhancement of individual harmonics, we note the difference between the phase-matching conditions for different parts of this harmonic. The phase mismatch varies as the laser pulse propagates through the plasma plume due to further ionization of the nonlinear medium. For harmonics in the plateau region, the phase mismatch attributed to free electrons is one to two orders of magnitude larger than the mismatch due to neutrals and singly charged ions. However, under resonance conditions, when the frequency of a part of a given harmonic becomes close to the frequency of inner-shell atomic transitions, the wave-number variation for this harmonic caused by atoms or ions might be significantly increased and the free electron effect may be cancelled [66]. Under these circumstances, it is possible to satisfy the optimal phase condition for a part of the single harmonic, with the consequent increase of conversion efficiency for this emission. The refractive index of the plasma in the vicinity of a resonant transition (λ_r) can be considerably changed thus allowing coincidence of the refractive

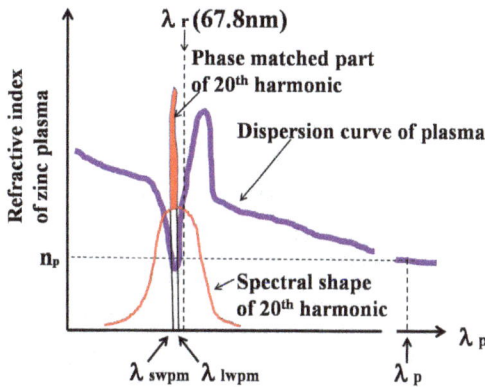

Fig. 3.20. Conditions of the phase matching between the waves of broadband pumps (1350 nm + 675 nm) and the 20th harmonic overlapping the resonance transition (λ_r = 67.8 nm) of the Zn plasma. Thick violet curve shows a dispersion of the refractive index of plasma. Thin red curve represents the spectral shape of the broadband 20th harmonic. The red filled area shows the enhancement of the part of this harmonic induced by the fulfillment of the phase matching conditions. Reproduced from [82] with permission from IOP Publishing.

indices of the plasma at the wavelengths of the pump and the part of harmonic emission (see the dispersion curve of plasma shown in Fig. 3.20). The short-wavelength wing of the resonance showing anomalous dispersion can create the conditions to satisfy this phase-matching condition. One can see that in the area marked by the λ_{swpm} and λ_{lwpm} the condition when the refractive index of plasma at the pump wavelength (n_p) becomes equal to the refractive index of plasma at the part of broadband harmonic. Here λ_{swpm} and λ_{lwpm} correspond to the short-wavelength and long-wavelength sides of the spectral area of phase matching.

Whether this mechanism affects (or does not affect) the phase relations between the interacting waves depends on many factors. Here we just proposed the explanation of the observed enhancement of relatively narrowband emission corresponding to the part of the whole spectral width of 20th harmonic. The joint influence of the processes at the micro-scale related with mechanisms described in the four-step model of resonance enhancement of harmonics [7] and the macroscopic processes related with the phase matching of the interacting waves can create the conditions for the generation of the intense emission possessing the same coherence properties as ordinary harmonics.

Another option to explain the observed enhanced emission line of zinc plasma upon excitation by two-color MIR + H2 multi-cycle pulses could be related with the lasing effect involving the ion transitions. One can assume that, analogously with the x-ray lasers, excitation of low-ionized plasma by ultrashort pulses increases the population of the discussed excited ion levels, causing stimulated emission at the corresponding wavelengths once population inversion is established between some ion levels. To support this assumption one has to analyze the $I_e(l \times g)$ dependencies, where I_e is the intensity of emitted radiation, l is the length, and g is the gain of the medium, as well as define the saturation conditions of the process. The difficulty in explaining the observed spectral peculiarities of the emission from Zn plasma using this approach is related with the unknown values of the lifetimes of the involved excited states and of the ratio between the absorption and the gain of the zinc plasma in this spectral range. Moreover, the polarization experiments do not support the assumption of lasing effect, since the gain of the medium should not depend so decisively on the polarization properties of the pump laser, as it was observed in [82]. In our case the rotation of pump polarization drastically changed the efficiency of both "non-resonant" and "resonant" harmonics. In particular, the 15° rotation of the quarter-wave plate placed on the path of pump radiation completely stopped the process of harmonic generation in Zn plasma. Finally, the enhancement of 20th harmonic corresponded to the classical quadratic dependence on the plasma length, once the sizes of plasma plume were changed between 0.5 and 5 mm. This quadratic dependence matches with the harmonic generation process rather than the lasing mechanism of enhanced emission.

The demonstrated method of optimization of the driving pulses wavelength for the resonance enhanced HHG opens new opportunities of harmonic generation of ultrashort pulses in the laser-produced plasma plumes and can be considered as a tool for various studies of different species (particularly large molecules and clusters) in the ablated conditions. The use of tunable two-color pump of LPP allows the analysis of the ionic states in plasmas, and broadens the subjects of studies compared with presently used HHG in gases. With these studies we demonstrate the application of MIR radiation for the analysis of the dynamics of nonlinear optical response of ablated solids compared with the 800-nm-class lasers commonly used for plasma HHG studies. The method allows a search for new opportunities in improvement of HHG conversion efficiency using the MIR laser sources, particularly using the multi-color pump of plasmas and optimization of single harmonic yield using tunable radiation.

Among the special features of HHG in laser-produced plasmas, we first of all note a wide range of medium characteristics available by varying the conditions of LPP. This applies to plasma parameters such as the plasma dimension, the density of ions, electrons, and neutral particles, and the degree of their excitation. The use of any elements of the periodic table that exist as solids largely extends the range of materials employed, together with thousands of complex solid-state samples, whereas only a few light rare gases are typically used in gas HHG schemes. The exploration of practically any available solid-state material through the nonlinear spectroscopy comprising laser ablation and harmonic generation can be considered as a new tool for material science.

Thus the plasma harmonic approach using tunable wavelength can be useful for producing an efficient source of short-wavelength ultra-short pulses for various applications and studies of the properties of harmonic emitters. This method of material science can be considered as one of the most important applications of HHG. Another application is the further improvement of the HHG efficiency through harmonic generation in specially prepared plasmas, which allow the spectral and structural studies of matter through the plasma harmonic spectroscopy.

3.4. Conclusions to Chapter 3

In this Chapter, we have discussed the advantages of the application of tunable mid-infrared pulses for the harmonic generation in the plasma media under the conditions of resonance-enhanced growth of single harmonics. High-order harmonics generated in laser-produced plasma can be resonantly enhanced in the energy range, which corresponds to the plasma ion transitions possessing large oscillator strengths. This effect was discovered using the fixed wavelength pump sources (Ti:sapphire laser). In order to optimize these enhancements as well as to study these ionic transitions in more detail we analyzed the application of the tunable mid-infrared radiation from optical parametric amplifier. The intensity of harmonics generated by such long fundamental wavelength (1250–1400 nm) was increased using the corresponding two-color laser field with the second harmonic field orthogonally polarized to the fundamental one.

For theoretical description of resonance-enhanced HHG a simple 1s–2p hydrogen-like atom model was used. This model cannot precisely describe the multielectron plasma ions used in the experiments, but it can serve for a qualitative explanation of the process. Though neither the present theory nor previous theories describe at the same time the enhancement of single harmonic, photo-induced decrease of neighboring harmonic, and second-harmonic induced growth of the whole yield of harmonic emission, the proposed approach allowed a qualitative analysis of the most important features of this phenomenon, namely the growth of harmonic yield for specific harmonic orders in each plasma medium under consideration during the wavelength tuning of the driving pulses near some ionic transitions.

The used laser and target material parameters are the same in theory and experiment. In this sense, the qualitative comparison of theory and experiment was clearly shown. As for comparison of the role of resonances, two graphs [Figs. 3.3(c) and 3.6] demonstrating the tuning of enhanced harmonics as well as the change of the order of this harmonic for different pumps were discussed. This approach allowed precise analysis of the role of those transitions by tuning various harmonics through the resonances under consideration. The enhancement of tunable harmonics in the regions of 27, 38, and 47 nm using tin, antimony, and chromium plasmas was demonstrated. These studies have shown that the application of tunable MIR pulses and their second harmonics can be used to optimize the earlier reported resonance enhancement of specific harmonics as well as to identify new spectral regions of harmonic enhancement.

The main scientific interest of the second research was about using the tunable pump to get a better understanding of the HHG at resonances of indium. To demonstrate this many individual spectra at different wavelengths of fundamental pump have been shown. The tunability was practically unlimited since the tuning of fundamental wavelength allowed the wavelength of high-order harmonic to be shifted over the wavelength of neighboring harmonic, which means the overlap of a full octave. It was shown that tuning of odd and even high-order harmonics along the strong resonance of laser-produced indium plasma using optical parametric amplifier of white-light continuum radiation (1250–1400 nm) allow different harmonics enhanced in the vicinity of the strong AIS of In II ions to be observed. Various peculiarities of this process were demonstrated and the theoretical model of the phenomenon of tunable harmonics enhancement in the region of 62 nm using indium plasma was discussed. The selective enhancement of the resonant harmonics in the two-color scheme has been confirmed by the theoretical model. This approach allowed the role of the AIS of In ions to be precisely analyzed by tuning various harmonics through the resonance under consideration. These studies show that the application of tunable MIR pulses can be used to optimize the earlier reported resonance enhancement of specific harmonics as well as to identify new areas of harmonic enhancement.

We also presented first observations of the zinc plasma harmonic spectra modification using two-color MIR pulses allowing definition of the excited neutral and ionic species responsible for the enhancement of single harmonics close to the resonances of medium. The discussed approach allowed the role of those transitions to be precisely analyzed by tuning various harmonics through the resonances under consideration at controllable conditions of excitation of neutral and ionic transitions. We analyzed the relation between the appearance of strong emission lines close to those transitions and the enhancement of specific harmonics. These studies allowed further development of the laser ablation induced HHG spectroscopy

160 *Interaction of MIR Radiation and Plasma*

of ablated species and showed that the application of tunable MIR pulses may both optimize earlier reported resonance enhancement of specific harmonics and identify new areas of harmonic enhancement. Micro-processes and propagation effect are discussed to describe the narrowing of the enhanced emission spectra from zinc plasma.

References

[1] R. A. Ganeev, M. Suzuki, T. Ozaki, M. Baba, and H. Kuroda, *Opt. Lett.* **31**, 1699 (2006).

[2] M. Suzuki, M. Baba, H. Kuroda, R. A. Ganeev, L. B. Elouga Bom, and T. Ozaki, *Opt. Express* **15**, 4112 (2007).

[3] M. Suzuki, M. Baba, R. A. Ganeev, H. Kuroda, and T. Ozaki, *J. Opt. Soc. Am. B* **24**, 2686 (2007).

[4] R. A. Ganeev, M. Suzuki, S. Yoneya, and H. Kuroda, *J. Appl. Phys.* **117**, 023114 (2015).

[5] D. B. Milošević, *J. Phys. B* **40**, 3367 (2007).

[6] D. B. Milošević, *Phys. Rev. A* **81**, 023802 (2010).

[7] V. Strelkov, *Phys. Rev. Lett.* **104**, 123901 (2010).

[8] M. V. Frolov, N. L. Manakov, and A. F. Starace, *Phys. Rev. A* **82**, 023424 (2010).

[9] M. Tudorovskaya and M. Lein, *Phys. Rev. A* **84**, 013430 (2011).

[10] V. V. Strelkov, M. A. Khokhlova, and N. Y. Shubin, *Phys. Rev. A* **89**, 053833 (2014).

[11] R. A. Ganeev, S. Odžak, D. B. Milošević, M. Suzuki, and S. H. Kuroda, *Laser Phys.* **26**, 075401 (2016).

[12] M. V. Frolov, N. L. Manakov, W.-H. Xiong, L.-Y. Peng, J. Burgdörfer, and A. F. Starace, *Phys. Rev. Lett.* **114**, 069301 (2015).

[13] I. J. Kim, C. M. Kim, H. T. Kim, G. H. Lee, Y. S. Lee, J. Y. Park, D. J. Cho, and C. H. Nam, *Phys. Rev. Lett.* **94**, 243901 (2005).

[14] L. Brugnera, D. J. Hoffmann, T. Siegel, F. Frank, A. Zair, J. W. G. Tisch, and J. P. Marangos, *Phys. Rev. Lett.* **107**, 153902 (2011).

[15] C. McGuinness, M. Martins, P. Wernet, B. F. Sonntag, P. van Kampen, J.-P. Mosnier, E. T. Kennedy, and J. T. Costello, *J. Phys. B* **32**, L583 (1999).

[16] R. D'Arcy, J. T. Costello, C. McGuinnes, and G. O'Sullivan, *J. Phys. B* **32**, 4859 (1999).

[17] C. McGuinness, M. Martins, P. van Kampen, J. Hirsch, E. T. Kennedy, J.-P. Mosnier, W. W. Whitty, and J. T. Costello, *J. Phys. B* **33**, 5077 (2000).

[18] G. Duffy, P. van Kampen, and P. Dunne, *J. Phys. B* **34**, 3171 (2001).

[19] J. B. West, J. E. Hansen, B. Kristensen, F. Folkmann, and H. Kjeldsen, *J. Phys. B* **36**, L327 (2003).

[20] A. Kramida, Y. Ralchenko, J. Reader, and NIST ASD Team (2013). *NIST Atomic Spectra Database* (ver. 5.1). National Institute of Standards and Technology, Gaithersburg, MD.

[21] R. A. Ganeev, M. Suzuki, M. Baba, and H. Kuroda, *Appl. Phys. Lett.* **86**, 131116 (2005).

[22] M. Suzuki, M. Baba, R. Ganeev, H. Kuroda, and T. Ozaki, *Opt. Lett.* **31**, 3306 (2006).

[23] R. A. Ganeev, V. V. Strelkov, C. Hutchison, A. Zaïr, D. Kilbane, M. A. Khokhlova, and J. P. Marangos, *Phys. Rev. A* **85**, 023832 (2012).

[24] M. Suzuki, M. Baba, H. Kuroda, R. A. Ganeev, and T. Ozaki, *Opt. Express* **15**, 1161 (2007).

[25] N. Rosenthal and G. Marcus, *Phys. Rev. Lett.* **115**, 133901 (2015).

[26] R. A. Ganeev, M. Suzuki, S. Yoneya, and H. Kuroda, *J. Appl. Phys.* **117**, 023114 (2015).

[27] D. B. Milošević, *J. Phys. B* **48**, 171001 (2015).

[28] D. B. Milošević, *J. Opt. Soc. Am. B* **23**, 308 (2006).

[29] R. A. Ganeev and D. B. Milošević, *J. Opt. Soc. Am. B* **25**, 1127 (2008).

[30] D. B. Milošević, W. Becker, and R. Kopold, *Phys. Rev. A* **61**, 063403 (2000).

[31] J. Rothhardt, S. Hädrich, S. Demmler, M. Krebs, S. Fritzsche, J. Limpert, and A. Tünnermann, *Phys. Rev. Lett.* **112**, 233002 (2014).

[32] R. A. Ganeev, M. Suzuki, M. Baba, and H. Kuroda, *J. Opt. Soc. Am. B* **23**, 1332 (2006).

[33] D. Popmintchev, C. Hernández-García, F. Dollar, C. Mancuso, J. A. Pérez-Hernández, M. C. Chen, A. Hankla, X. Gao, B. Shim, A. L. Gaeta, M. Tarazkar, D. A. Romanov, R. J. Levis, J. A. Gaffney, M. Foord, S. B. Libby, A. Jaron-Becker, A. Becker, L. Plaja, M. M. Murnane, H. C. Kapteyn, and T. Popmintchev, *Science* **350**, 1225 (2015).

[34] A.V. Andreev, R. A. Ganeev, H. Kuroda, S. Yu. Stremoukhov, and O. A. Shoutova, *Eur. Phys. J. D* **67**, 22 (2013).

[35] P. B. Corkum and F. Krausz, *Nat. Phys.* **3**, 381 (2007).

[36] J. P. Marangos, S. Baker, N. Kajumba, J. S. Robinson, J. W. G. Tisch, and R. Torres, *Phys. Chem. Chem. Phys.* **10**, 35 (2008).

[37] C. Jin, H. J. Wörner, V. Tosa, A.-T. Le, J. B. Bertrand, R. R. Lucchese, P. B. Corkum, D. M. Villeneuve, and C. D. Lin, *J. Phys. B: At. Mol. Opt. Phys.* **44**, 095601 (2011).

[38] K. J. Schafer, B. Yang, L. F. DiMauro, and K. C. Kulander, *Phys. Rev. Lett.* **70**, 1599 (1993).

[39] P. B. Corkum, *Phys. Rev. Lett.* **71**, 1994 (1993).

[40] J. Levesque, D. Zeidler, J. P. Marangos, P. B. Corkum, and D. M. Villeneuve, *Phys. Rev. Lett.* **98**, 183903 (2007).

[41] R. A. Ganeev, H. Singhal, P. A. Naik, V. Arora, U. Chakravarty, J. A. Chakera, R. A. Khan, P. V. Redkin, M. Raghuramaiah, and P. D. Gupta, *J. Opt. Soc. Am. B* **23**, 2535 (2006).

[42] D. Shiner, B. E. Schmidt, C. Trallero-Herrero, H. J. Wörner, S. Patchkovskii, P. B. Corkum, J.-C. Kieffer, F. Légaré and D. M. Villeneuve, *Nature Phys.* **7**, 464 (2011).

[43] U. Fano, *Phys. Rev.* **124**, 1866 (1961).

[44] F. Keller and H. Lefebvre-Brion, *Z. Phys. D: At. Mol. Clusters* **4**, 15 (1986).

[45] M. Y. Amusia and J.-P. Connerade, *Rep. Prog. Phys.* **63**, 41 (2000).

[46] J. F. Reintjes, *Nonlinear Optical Parametric Processes in Liquids and Gases*, Academic (1984).

[47] E. S. Toma, P. Antoine, A. de Bohan, and H. G. Muller, *J. Phys. B* **32**, 5843 (1999).

[48] M. B. Gaarde and K. J. Schafer, *Phys. Rev. A* **64**, 013820 (2001).

[49] R. Bartels, S. Backus, E. Zeek, L. Misoguti, G. Vdovin, I. P. Christov, M. M. Murnane and H. C. Kapteyn, *Nature* **406**, 164 (2000).

[50] R. Taïeb, V. Véniard, J. Wassaf, and A. Maquet, *Phys. Rev. A* **68**, 033403 (2003).

[51] V. Redkin and R. A. Ganeev, *Phys. Rev. A* **81**, 063825 (2010).

[52] R. A. Ganeev, Z. Wang, P. Lan, P. Lu, M. Suzuki, and H. Kuroda, *Phys. Rev. A* **93**, 043848 (2016).

[53] P. Lan, E. J. Takahashi, and K. Midorikawa, *Phys. Rev. A* **81**, 061802 (2010).

[54] E. Cormier and M. Lewenstein, *Eur. Phys. J. D* **12**, 227 (2000).

[55] J. Mauritsson, P. Johnsson, E. Gustafsson, A. L'Huillier, K. J. Schafer, and M. B. Gaarde, *Phys. Rev. Lett.* **97**, 013001 (2006).

[56] T. Pfeifer, L. Gallmann, M. J. Abel, D. M. Neumark, and S. R. Leone, *Opt. Lett.* **31**, 975 (2006).

[57] Y. Yu, X. Song, Y. Fu, R. Li, Y. Cheng, and Z. Xu, *Opt. Express* **16**, 686 (2008).

[58] X.-S. Liu and N.-N. Li, *J. Phys. B: At. Mol. Opt. Phys.* **41**, 015602 (2008).

[59] D. Charalambidis, P. Tzallas, E. P. Benis, E. Skantzakis, G. Maravelias, L. A. A. Nikolopoulos, A. P. Conde, and G. D. Tsakiris, *New J. Phys.* **10**, 025018 (2008).

[60] I.J. Kim, G. H. Lee, S. B. Park, Y. S. Lee, T. K. Kim, C. H. Nam, T. Mocek and K. Jakubczak, *Appl. Phys. Lett.* **92**, 021125 (2008).

[61] H. R. Telle, G. Steinmeyer, A. E. Dunlop, J. Stenger, D. H. Sutter, and U. Keller, *Appl. Phys. B* **69**, 327 (1999).

[62] V. Tosa, E. Takahashi, Y. Nabekawa, and K. Midorikawa, *Phys. Rev. A* **67**, 063817 (2003).

[63] H. R. Lange, A. Chiron, J. F. Ripoche, A. Mysyrowicz, P. Breger, and P. Agostini, *Phys. Rev. Lett.* **81**, 1611 (1998).

[64] R. A. Ganeev, M. Suzuki, and H. Kuroda, *Phys. Rev. A* **89**, 033821 (2014).

[65] R. A. Ganeev, V. Tosa, K. Kovács, M. Suzuki, S. Yoneya and H. Kuroda, *Phys. Rev. A* **91**, 043823 (2015).

[66] R. A. Ganeev, T. Witting, C. Hutchison, V. V. Strelkov, F. Frank, M. Castillejo, I. Lopez-Quintas, Z. Abdelrahman, J. W. G. Tisch, and J. P. Marangos, *Phys. Rev. A* **88**, 033838 (2013).

[67] J. L. Krause, K. J. Schafer, and K. C. Kulander, *Phys. Rev. A* **45**, 4998 (1992).

[68] M. Protopapas, C. H. Keitel, and P. L. Knight, *Rep. Prog. Phys.* **60**, 389 (1997).

[69] A. Castro, H. Appel, M. Oliveira, C. A. Rozzi, X. Andrade, F. Lorenzen, M. A. L. Marques, E. K. U. Gross, and A. Rubio, *Phys. Stat. Sol. B* **243**, 2465 (2006).

[70] K. Burnett, V. C. Reed, J. Cooper, and P. L. Knight, *Phys. Rev. A* **45**, 3347 (1992).

[71] N. Rosenthal and G. Marcus, *Phys. Rev. Lett.* **115**, 133901 (2015).

[72] C. Hutchison, R. A. Ganeev, M. Castillejo, I. Lopez-Quintas, A. Zair, S. J. Weber, F. McGrath, Z. Abdelrahman, M. Oppermann, M. Martín, D. Y. Lei, S. A. Maier, J. W. Tisch, and J. P. Marangos, *Phys. Chem. Chem. Phys.* **15**, 12308 (2013).

[73] H. Hora, *Plasmas at High Temperature and Density*, Springer, Heidelberg (1991).

[74] B. Rus, P. Zeitoun, T. Mosek, S. Sebban, M. Kalal, A. Demir, G. Jamelot, A. Klisnick, B. Kralikova, J. Skala, and G. J. Tallents, *Phys. Rev. A* **56**, 4229 (1997).

[75] R. A. Ganeev, M. Suzuki, M. Baba, and H. Kuroda, *Opt. Spectrosc.* **99**, 1000 (2005).

[76] R. A. Ganeev, H. Singhal, P. A. Naik, J. A. Chakera, H. S. Vora, R. A. Khan, and P. D. Gupta, *Phys. Rev. A* **82**, 053831 (2010).

[77] R. A. Ganeev, C. Hutchison, A. Zaïr, T. Witting, F. Frank, W. A. Okell, J. W. G. Tisch, and J. P. Marangos, *Opt. Express* **20**, 90 (2012).

[78] R. A. Ganeev, H. Singhal, P. A. Naik, V. Arora, U. Chakravarty, J. A. Chakera, R. A. Khan, I. A. Kulagin, P. V. Redkin, M. Raghuramaiah, and P. D. Gupta, *Phys. Rev. A* **74**, 063824 (2006).

[79] R. A. Ganeev, L. B. Elouga Bom, J.-C. Kieffer, and T. Ozaki, *Phys. Rev. A* **75**, 063806 (2007).

[80] R. A. Ganeev, J. Zheng, M. Wöstmann, H. Witte, P. V. Redkin, and H. Zacharias, *Eur. Phys. J. D* **68**, 325 (2014).

[81] G. Duffy and P. Dunne, *J. Phys. B* **34**, L173 (2001).

[82] R. A. Ganeev, M. Suzuki, S. Yoneya, V. V. Strelkov, and H. Kuroda, *J. Phys. B: At. Mol. Opt. Phys.* **49**, 055402 (2016).

[83] R. A. Ganeev, Z. Abdelrahman, F. Frank, T. Witting, W. A. Okell, D. Fabris, C. Hutchison, J. P. Marangos, and J. W. G. Tisch, *Appl. Phys. Lett.* **104**, 021122 (2014).

[84] R. A. Ganeev, M. Baba, M. Suzuki, S. Yoneya, and H. Kuroda, *J. Appl. Phys.* **116**, 243102 (2014).

[85] R. A. Ganeev, H. Singhal, P. A. Naik, I. A. Kulagin, P. V. Redkin, J. A. Chakera, M. Tayyab, R. A. Khan, and P. D. Gupta, *Phys. Rev. A* **80**, 033845 (2009).

[86] A. M. Crooker and K. A. Dick, *Canad. J. Phys.* **46**, 1241 (1968).

[87] K. A. Dick, *Canad. J. Phys.* **46**, 1291 (1968).

[88] C. G. Back, M. D. White, V. Pejčev, and K. J. Ross, *J. Phys. B* **14**, 1497 (1981).

[89] M. W. D. Mansfield, *J. Phys. B* **14**, 2781 (1981).

[90] N. L. S. Martin, *J. Phys. B* **17**, 1797 (1984).

[91] K. Sommer, M. A. Baig, and J. Hormes, *Z. Phys. D* **4**, 313 (1987).

[92] B. Predojević, D. Šević, V. Pejčev, B. P. Marinković, and D. M. Filipović, *J. Phys. B* **36**, 2371 (2003).

[93] G. V. Marr and J. M. Austin, *J. Phys. B* **2**, 107 (1969).

[94] D. C. Yost, T. R. Schibli, J. Ye, J. L. Tate, J. Hostetter, M. B. Gaarde, and K. J. Schafer, *Nature Phys.* **5**, 815 (2009).

[95] W.-H. Xiong, J.-W. Geng, J.-Y. Tang, L.-Y. Peng, and Q. Gong, *Phys. Rev. Lett.* **112**, 233001 (2014).

Chapter 4

Quasi-phase-matching in Plasmas
Using Mid-infrared Pulses

In this Chapter, we demonstrate the quasi-phase-matching of a group
of harmonics generated in Ag multi-jet plasma using tunable pulses in
the region of 1160–1540 nm and their second harmonic emission. The
numerical treatment of this effect includes microscopic description of
the harmonic generation, propagation of the pump pulse, and the
propagation of the generated harmonics. A few tens fold growth
of harmonics under the conditions of quasi-phase-matching in the
region of 35 nm using eight-jet plasma compared with the case of
imperforated plasma was achieved.

Further, we analyze the spatial conditions of the quasi-phase-
matching in the multi-jet laser-produced silver plasma. The studies of
the off-axis and on-axis spatial components of harmonics allowed the
demonstration of significant enhancement of a group of harmonics
in the latter case. The divergence of enhanced radiation was a
few times lower compared with the case of harmonic generation
in the extended homogeneous plasma. We show the appearance of
plasma emission, together with phase-mismatch, that deteriorate
the conditions of high-order harmonic generation due to significant
phase distortion between the interacting waves. These two-color-
pump-induced HHG under the conditions of multi-jet plasma showed
the complexity of the phase relations between the two orthogonally
polarized pumps and harmonic wave, which led to variable conditions

of the quasi-phase-matching (QPM) in the cases of shorter- and longer-wavelength pumps.

Finally, we compare the resonance-induced enhancement of single harmonic and the quasi-phase-matching-induced enhancement of the group of harmonics during propagation of the tunable mid-infrared femtosecond pulses through the perforated laser-produced indium plasma. We show that the enhancement of harmonics using the macro-process of quasi-phase-matching is comparable with the one using micro-process of resonantly enhanced harmonic. These studies demonstrate that joint implementation of the two methods to increase harmonic yield could be a useful tool for generation of strong short-wavelength radiation in different spectral regions. We compare these effects in indium, as well as in other plasmas.

4.1. Application of Mid-infrared Pulses for Quasi-phase-matching of High-order Harmonics in Silver Plasma

4.1.1. *Early studies of quasi-phase-matching*

The improvement of high-order harmonic generation efficiency through the amendment of phase matching conditions [1] in gaseous and plasma media during propagation of ultrashort laser plasma is aimed at formation of reliable sources of coherent extreme ultraviolet radiation for various applications in biomedicine, physical and chemical experiments, spectroscopy, etc. The QPM of the interacting waves is a reliable method for the enhancement of harmonic yield [2–13]. The QPM is aimed at improving the relative phase between pump and harmonic waves. Once the envelopes of the pulses of these two waves become overlapped over the whole range of medium the emission of short-wavelength photons from each emitter accumulates and quadratically increases with medium length. One of methods to fulfill QPM is a modulation of active medium density. In particular, modification of extended medium onto a group of separated gas or plasma jets allows formation of conditions when, for some group of harmonics, the relative phases of pump and harmonic waves become maintained along the whole set of media bunches. As it has been shown in multi-jet gases [4,5,8,9,13], once the waves depart from the medium their relative phase flips. After that the process of frequency conversion can be efficiently maintained in another bunch of medium. Plasma jet schemes also allow such enhancement [14,15]. In addition, HHG using the QPM in density-modulated nanoparticle composite was recently reported [16].

There are plenty of the methods introduced during studies of gas and plasma harmonics to improve HHG conversion efficiency. In the case of the harmonics generated in the laser-produced plasma plumes, those include the application of nanoparticles and clusters as the harmonic emitters, the commensurate and incommensurate two-color pump of plasmas, the application of extended plasmas, the resonance enhancement of single harmonic, and the quasi-phase-matching of harmonics. Among them the QPM and two-color pump approaches have shown to be an attractive way in improving harmonic yield in different spectral ranges. The combination of these two methods allowed the amendment of harmonics during HHG experiments in laser-produced plasmas [17]. Those and most other refereed studies were carried out using the conventional Ti:sapphire lasers and their second harmonics.

In the meantime, the application of longer wavelength sources may lead to further growth of harmonics in the multi-jet plasmas compared with the imperforated media. The attractiveness of using the longer-wavelength sources for the amendment of plasma HHG has been recently revealed during experiments with nanoparticles [18]. One can assume that tunable optical parametric amplifiers and their second harmonics are the advanced choice for pumping the modulated LPP to maintain efficient QPM conditions.

In this section, we show the quasi-phase-matching of a group of harmonics generated in the multi-jet plasma produced by laser ablation of bulk silver target using tunable pulses in the mid-infrared region of 1250–1400 nm and their second harmonic emission. We analyze the observation of 12-fold growth of shorter-wavelength harmonics compared with longer wavelength ones in the eight-jet plasma and more than 60× growth of the 39th harmonic generated in the multi-jet LPP compared with the imperforated extended plasma. We also show the numerical treatment of this effect, which includes the microscopic description of harmonic generation, as well as the macroscopic consideration of HHG including the propagation of pump and harmonic pulses through the medium [19].

4.1.2. *Experimental conditions and HHG in silver plasma plumes using tunable 1250–1400 nm, 70 fs pulses under the conditions of quasi-phase-matching*

The experimental setup consisted on three parts: (a) Ti:sapphire laser, (b) travelling-wave OPA of white-light continuum, and (c) setup for high-order harmonic generation using propagation of amplified signal pulse of parametric amplifier through the extended imperforated and perforated LPP. The details of the laser and experimental conditions [Fig. 4.1(a)] were described in section 3.2. The signal radiation from OPA was used for HHG in LPP. The intensity of 1310-nm pulses focused by 400-mm focal length lens inside the extended plasma was $2 \times 10^{14} \, \mathrm{W \, cm^{-2}}$. This driving pulse was focused into the prepared extended plasma at a distance of $\sim 100 \, \mu \mathrm{m}$ above the target surface. Most of experiments were carried out using the two-color pump of LPP.

Silver was used as the target for ablation. The size of the target where the ablation occurred was 5 mm. The plasma sizes were $5 \times 0.08 \, \mathrm{mm^2}$. To create multi-jet plasmas a multi-slit mask (MSM) was used. The size of the slits was 0.3 mm with distance between them also 0.3 mm [Fig. 4.1(b)]. The MSM was installed between the focusing cylindrical lens and target such as to divide the continuous 5-mm-long plasma in ~ 8 times 0.3 mm plasma jets with ~ 0.3 mm separation. The number of plasma jets can be increased by tilting the MSM, as shown in Fig. 4.1(b). In particular, tilting the mask at 45° allowed the formation of 11 jets.

Below we analyze the reported data of the QPM of the groups of harmonics using the MIR pulses and their second harmonics (1310 nm + 655 nm) [19]. The principles of QPM in LPP have previously been demonstrated using the 800 nm lasers. Here we discuss

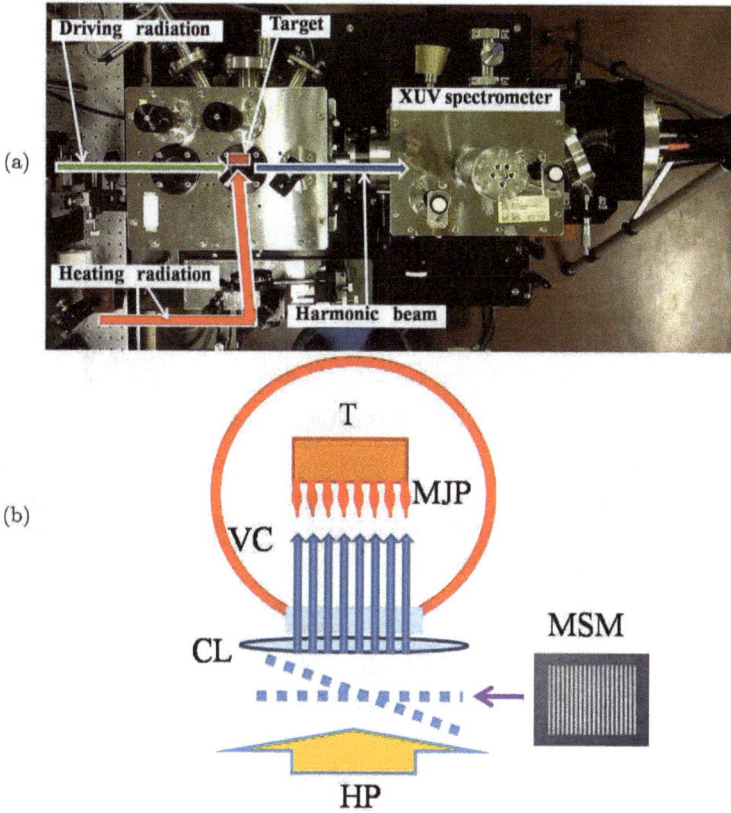

Fig. 4.1. (a) Experimental setup for harmonic generation in LPP. (b) Multi-jet plasma formation on the surface of silver target. HP, heating picosecond pulse from Ti:sapphire laser; CL, cylindrical lens; VC, vacuum chamber; T, silver target; MSM, multi-slit mask; MJP, multi-jet plasma. Reproduced from [19]with permission from Optical Society of America.

the stronger enhancement of the group of harmonics around 39th harmonic (H39), which could be achieved in the case of MIR sources and two-color pump approach.

Figure 4.2 shows two harmonic spectra using the two-color pump of the silver plasma produced using the fluence of heating pulse of $F = 1.0\,\mathrm{J\ cm^{-2}}$. The thick red curve shows the harmonic spectrum obtained in the extended imperforated 5-mm-long plasma. A featureless spectrum of gradually decreased harmonics starting

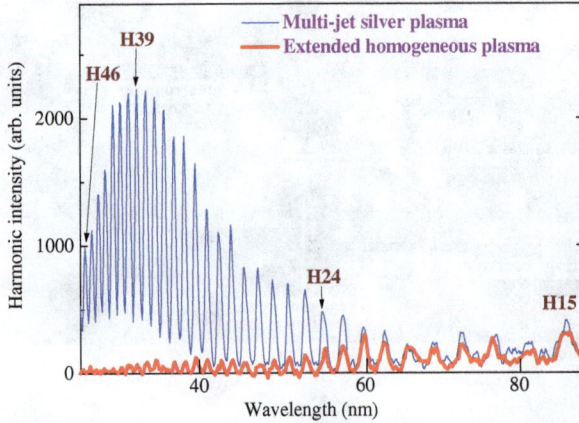

Fig. 4.2. Harmonic spectra from the extended homogeneous plasma (thick red curve) and multi-jet plasma (thin blue curve) produced on the silver target using the two-color pump (1310 nm+655 nm). Reproduced from [19] with permission from Optical Society of America.

from the 15th order up to the 46th order just shows the conventional plateaulike distribution of gradually decreasing harmonics. Once an extended plasma was separated on a group of jets by using the MSM placed in front of ablating target, a significant variation of harmonic distribution was observed (thin blue curve). A group of harmonics centered near the H39 was notably enhanced compared with the lower orders. The enhancement factor $\sim 12 \times$ was achieved for the maximally enhanced harmonics compared with the lower-order ones. Furthermore, more than $30 \times$ growth of the 39th harmonic generated in multi-jet LPP compared with extended plasma was achieved. The order of harmonics which satisfied the condition of QPM was larger compared with the case of using 800 nm pump due to lesser dispersion of plasma in the MIR range. The maximally enhanced harmonic order (q_{qpm}) corresponded to the relation $q_{qpm} = 1.1 \times 10^{18}/(l_{jet} \times N_e)$, where l_{jet} and N_e are the length of single plasma jet in multi-jet plume (measured in mm) and electron density of plasma (measured in cm^{-3}) respectively. Note that the size of each jet was ~ 0.3 mm.

The definition of maximally enhanced harmonic goes from the definition of the coherence length of this harmonic. More details can be found in [20, 21]. Briefly, the plasma dispersion-induced phase mismatch for the qth harmonic is defined by the relation $\Delta k_{disp} = qN_ee^2\lambda/4\pi m_e\varepsilon_0 c^2$, where λ is the wavelength of driving radiation, N_e, m_e and e are the density, the mass and the charge of the electron, c is the light velocity, and ε_0 is the vacuum permittivity [22]. The coherence length of this harmonic is $L_{coh} = \pi/\Delta k \approx \pi/\Delta k_{disp} = 4\pi^2 m_e\varepsilon_0 c^2/qN_ee^2\lambda$. From this expression the coherence length (in mm) under the conditions of using the 1300 nm driving laser could be presented as $L_{coh} \approx 1.1 \times 10^{18}/(N_e \times q_{qpm})$. This simple formula is useful since it allows the electron density to be defined by knowing the coherence length, which is in fact the size of single plasma jet under the conditions of the QPM in the multi-jet structure, and the maximally enhanced harmonic order. The decrease of heating pulse fluence on the surface of ablating target should lead to a decrease of electron density, due to less amount of ablation- and tunnel-induced electrons, in the plasma plume followed by the shift of q_{qpm} towards the shorter-wavelength region. Similarly, one can anticipate that, for the plasma jets of different sizes, the maximally enhanced harmonics will also be tuned along the XUV spectrum. The relation $q_{qpm} = 1.1 \times 10^{18}/(l_{jet} \times N_e)$ was taken from the above formula assuming the equality of the length of single jet (l_{jet}) and the coherence length of the corresponding harmonic, which had the highest QPM-induced enhancement.

The ablation using larger fluence of heating pulse ($F = 1.2\,\mathrm{J\,cm^{-2}}$) led to some change of relative intensities of similar harmonics in the cases of imperforated and perforated LPP. The spectral distribution of harmonics generated in the extended 5-mm-long Ag plasma is shown in the upper panel of Fig. 4.3. Installation of MSM orthogonally to the optical axis of heating beam led to dramatic enhancement of the harmonics around the H30 (middle panel), similarly to previous case. However, one can see a difference in maximally enhanced harmonics in this and previous cases (H30 and H39). This difference was caused by different concentration of free electrons in the plasma area in these two cases (i.e. at the heating fluencies of 1.2 and 1.0 J cm^{-2}). The tilting of MSM at angle 35° increased the number of jets from 8 to 10. A decrease

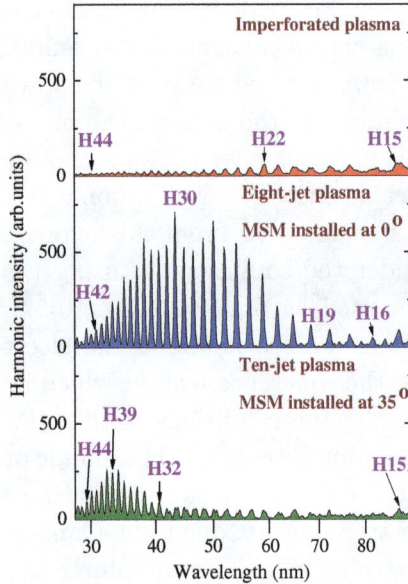

Fig. 4.3. Harmonic spectra from silver plasma using two-color pump (1310 + 655 nm). (upper panel) Extended imperforated 5-mm-long plasma plume. (middle panel) Eight-jet plasma produced by installation of multi-slit mask orthogonally to the optical axis of propagation of the heating pulse towards the target. (bottom panel) Tilting of MSM at 35° with regard to the axis of heating beam propagation. In that case, the plasma contained ten jets. The size of single jet was 0.25 mm. Reproduced from [19] with permission from Optical Society of America.

of single jet sizes led to increase of q_{qpm} in accordance with the above relation (from H30 to H39). The maximally enhanced group of harmonics was tuned towards the shorter wavelength range (bottom panel). One can compare the intensities of those harmonics with the intensities of similar orders in the case of imperforated plasma plume (upper panel). The ratio of these intensities shows the enhancement factor of harmonic emission achieved in 10-jet plasma, which was approximately 40× in the region of 34 nm.

4.1.3. *Theory of QPM*

The numerical treatment of the problem under consideration included three components: microscopic description of the HHG process, propagation of the pump pulse, and the propagation of the generated harmonics. The harmonic generation and the corresponding harmonic dipole moment were modeled using the strong-field formalism developed by Lewenstein *et al* [23], using the following expression for the dipole moment:

$$d_{HH}(t) = \frac{ie}{2\omega_o^{5/2} m_e} \int_{-\infty}^{t} \left(\frac{\pi}{\varepsilon + i(t - t_s)} \right)^{\frac{3}{2}} d(p_{st} - eA(t_s))d^*$$

$$\times (p_{st} - eA(t))E(t_s)e^{\frac{-iS(t,t_s)}{\hbar}} dt_s + c.c., \qquad (4.1)$$

where ω_0 is the pump carrier frequency, $d(p)$ is the dipole moment, p_{st} is the saddle-point canonical momentum, $A(t)$ is the vector-potential of the electric field $E(t)$, and $S(t, t_s) = \int_{ts}^{t} (l_p - \frac{1}{2}[p - eA(t')]^2/m_e)dt'$ is the classical action with l_p being the ionization potential. No approximation of zero velocity of the ionized electron was used, which allowed for the accurate description of the spectra also well below the harmonic threshold. Analytic expressions for dipole transition moments of the hydrogen-like atoms $d(p) = i2^{7.25}[\hbar\omega_0 m_e^2 l_p]^{1.25}\pi^{-1}p/(p^2 + \alpha)^3$ with $\alpha = 2\sqrt{m_e l_p}$ were utilized, and ADK formalism was used to describe the ionization rate:

$$\frac{\partial \rho}{\partial t} - \left(\frac{3m_e c^4}{\pi \hbar^3 n^{4,5}} \right) \left(\frac{4}{\varepsilon n^4} \right)^{2n.-15} e^{-\frac{2}{3n^3 \varepsilon}} \qquad (4.2)$$

with ε being the electric field in the atomic units and $n_s = \sqrt{2l_p \hbar^2/(m_e e^4)}$.

For the propagation of the pump pulse a (1+1)D unidirectional propagation equation was numerically solved without relying on the slowly-varying envelope approximation, which allowed both pulses to be included in the same formalism:

$$\frac{\partial E(z, \omega)}{\partial z} = i\beta(z, \omega)E(z, \omega) + \frac{i\omega^2}{2c^2 \varepsilon_0 \beta(z, \omega)} P_{NL}(z, \omega) \qquad (4.3)$$

where $\beta(z, \omega) = \omega n(z, \omega)/c$ is the wavenumber and the $P_{NL}(z, \omega)$ is the nonlinear polarization, determined by

$$P_{NL}(z, t) = \varepsilon_0 \chi_3 E(z, t) - e\rho d_e - 2l_p \int_{-\infty}^{t} \frac{1}{E(z, t')} \frac{\partial \rho(z, t')}{\partial t} dt' \qquad (4.4)$$

where χ_3 is the third-order susceptibility and d_e is the electron displacement with $d_R = -eE(z,t)/m_e$.

The defocusing of the pulse can be ignored at the considered propagation lengths, as they are shorter than the Rayleigh length. The contribution to the phase mismatch of both linear refractive index of silver plasma and the ionized electrons to the refractive index, together with corresponding dispersion terms to all orders, were included in the simulation. In addition, loss of the pump beam due to ionization as well as the Kerr nonlinearity was included in the simulation, though these two effects do not play a major role in the pulse evolution. For the details of the numerical treatment see [16]. Finally, the propagation of the high-order harmonics is governed by the (1+1)D propagation equation, which incorporates a high-harmonic polarization $P_{HH} = Ned_{HH}$ using harmonic source dipole moment d_{HH} calculated as described above, as well as harmonic reabsorption by the silver plasma.

In Fig. 4.4(a), the map of the spectrum evolution is presented for parameters reproducing the parameters of experiment shown in Fig. 4.3, as detailed in the caption. One can see that for most harmonic numbers, periodic modulation of the intensity is predicted, and the output power remains low. However, in the spectral range around the 19th harmonic the intensity steadily increases, as a result of the quasi-phase-matching achieved by the modulation of plasma density with a whole period of 0.6 mm, which contained a filled zone (0.3 mm) and empty zone (0.3 mm). Roughly 11 periods of the modulation fit in the propagation length, which would theoretically correspond to roughly two orders of magnitude enhancement of the harmonic intensity at the quasi-phase-matched number in comparison to non-quasi-phase-matched harmonic numbers. In the simulation, we achieve a factor of roughly 60. Note that both even and odd harmonics are generated due to two-color excitation. In comparison to the experiment, the QPM was predicted at the harmonic order of 19 while relatively broad range of QPM around harmonic order of 30 is observed in the experiment (Fig. 4.3,

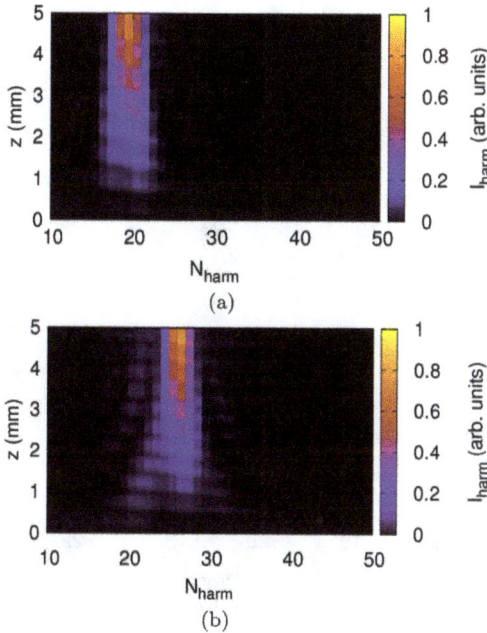

Fig. 4.4. (a) The map of the harmonic spectrum as a function of the propagation length for plasma density modulated with period of 0.6 mm. The two-color 70-fs pump pulses are considered, with peak intensities of $2 \times 10^{14}\,\mathrm{W\,cm^{-2}}$ and $0.54 \times 10^{14}\,\mathrm{W\,cm^{-2}}$, centered at 1310 and 665 nm, respectively. The silver plasma with density of $9.4 \times 10^{16}\,\mathrm{cm^{-3}}$ and density modulation depth of 0.9 is assumed. (b) The map of the harmonic spectrum as a function of the propagation length for plasma density modulated with period of 0.45 mm. Other parameters are the same as in Fig. 4.4(a). Reproduced from [19] with permission from Optical Society of America.

second panel). The difference can be explained by the contribution of the plasma to the index of refraction, which rather sensitively depends on the pump field and makes accurate reproduction of the numerical conditions somewhat challenging.

To exemplify the sensitive dependence of the generated spectra on the period of the modulation, the simulations were repeated with a modulation period of 0.45 mm, as shown in Fig. 4.4(b). One can see that the maximum of the spectrum has shifted to

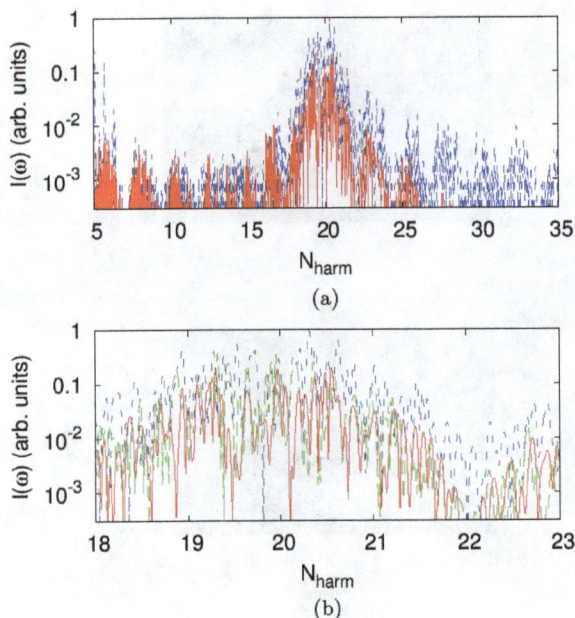

Fig. 4.5. The output spectra with (blue dashed) and without (red solid) second-harmonic pump pulse. In (a), the overview spectrum is shown, while in (b) the zoom-in around the quasi-phase-matched frequency is demonstrated. In (b), the green short-dashed curve is for the second harmonic conversion efficiency of 6.5%. Other parameters are as in Fig. 4.4. Reproduced from [19] with permission from Optical Society of America.

the harmonic order of 26. This result is in accordance with the relation that the quasi-phase-matched harmonic number is inversely proportional to the period, with expected harmonic order of 25. This offers a possibility to control the harmonic spectrum by changing the modulation by, e.g., tilting the mask, as explained in the experimental part of this chapter.

The output spectra are presented in Fig. 4.5 for both single-color (1310 nm) and two-color pumps. By adding second-harmonic pump, roughly one order of magnitude increase of the harmonic amplitude in the QPM region is achieved. Note that, as one can see in Fig. 4.5(a), the intensity contrast is much less pronounced outside of the quasi-phase-matched region, which indicates the intricate role of the second harmonic pump in the HHG process. On the

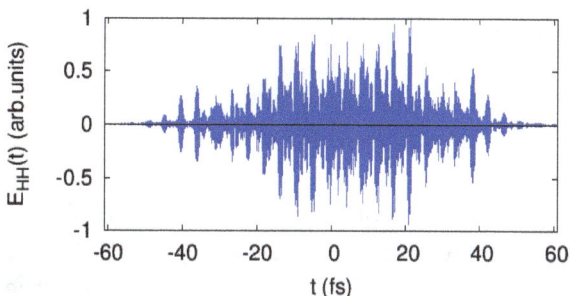

Fig. 4.6. The output temporal profile of the harmonics in the spectral range from 17 to 23 harmonics for the parameters considered in Fig. 4.4. Reproduced from [19] with permission from Optical Society of America.

other hand, the difference of efficiencies is quite homogeneous in the quasi-phase-matched region as shown in Fig. 4.5(b) except for fast modulation. Also, one can note that even a second-harmonic generated with a smaller efficiency of 6.5%, as presented by the green short-dashed curve in Fig. 4.5(b), is sufficient to notably increase the efficiency of HHG.

Finally, in Fig. 4.6 the temporal profile of the QPM harmonics in the spectral range from 17th to 23rd harmonic for the parameters considered in Fig. 4.4 is presented. One can see a periodic pattern in the temporal profile. This pattern is probably related to the phase mismatch varying along the temporal axis due to the variation of the electron density created by the pump pulse and corresponding modulation of the phase matching conditions and the phase accumulated during propagation. In general, absence of maximum at some time moment favored by the quasi-phase-matching is predicted.

4.2. On- and Off-axis Quasi-phase-matching of the Harmonics Generated in the Multi-jet Laser-produced Plasmas

4.2.1. *Description of the problem*

The QPM in the laser-produced plasmas has shown the attractive features comprising both the advantages realized during QPM in gases and the peculiarities of plasma HHG [14,15]. The optimization of multi-jet formation for HHG requires the manipulation of the relative phases of driving and harmonic waves. As already discussed in Chapter 2, one of the ways to change a dephasing between these waves is by applying additional driving field. Application of the two fields commonly comprising of the radiation of Ti:sapphire laser and its second harmonic has long been considered as one of advanced methods for the enhancement of gas HHG [24,25]. It would be interesting to analyze the correlation between the phases of two pumps and harmonics under the conditions of spatial QPM in the multi-jet plasma, as well as compare the two-color pump (TCP) and single-color pump (SCP) at different conditions of such experiments.

In this section, we analyze the spatial characteristics of QPM in the LPP. The studies of the off-axis and on-axis components of harmonics allow the demonstration of significant enhancement of a group of harmonics in the latter case. This TCP-induced HHG under the conditions of multi-jet plasma shows the complexity of the phase relations between the two orthogonally polarized pumps and harmonic wave, which leads to less favorable conditions for the QPM compared with the SCP [26].

A scheme similar to previous MIR HHG studies was used in these investigations excluding the driving source [806 nm instead of MIR (1200–1600 nm) pulses]. The uncompressed radiation of the Ti:sapphire laser operated at 10 Hz pulse repetition rate was used as a heating pulse (central wavelength $\lambda = 806$ nm, pulse duration 370 ps, pulse energy up to $E_{hp} = 4$ mJ) for extended plasma formation. The heating pulse was focused using a 200 mm focal-length cylindrical lens inside the vacuum chamber containing an ablating target to create the extended plasma plume above the target surface. The focusing of the heating pulse on the target surface produced the extended imperforated plasma. The intensity of the heating pulses on a plain target surface was varied up to 5×10^9 W cm^{-2}. The compressed driving pulse from the same laser with the energy of up to $E_{dp} = 5$ mJ and 64 fs pulse duration was used, after 45 ns from the beginning of ablation, for the harmonic generation in the plasma plume. The laser pulse was close to the transform-limited one. The pulse duration was measured using the autocorrelation technique. The driving pulse was focused using a 400 mm focal-length spherical lens onto the prepared plasma from perpendicular direction, at a distance of $\sim$100 μ m above the target surface. The confocal parameter of the focused driving beam was 18 mm. The intensity of the driving pulse in the focal area was varied up to 9×10^{14} W cm^{-2}.

Silver was used as the target for ablation. The size of the target where the ablation occurred was 5 mm. To create multi-jet plasma two multi-slit masks were used. The sizes of the slits of these masks were 0.3 and 0.8 mm with the distance between them 0.3 and 0.8 mm respectively. The first MSM was installed between the focusing cylindrical lens and target such as to divide the continuous 5-mm plasma in eight 0.3-mm-long plasma jets with $\sim$0.3 mm separation. The image of these jets was captured from the top of the vacuum chamber and is shown in Fig. 4.7. In the experiments, both imperforated and perforated plasmas were used as the nonlinear optical media. Here "imperforated plasma" refers to a plasma formed during the irradiation of the surface of ablating target by the unmasked heating pulses. In that case one can produce the extended homogeneous plasma along the target surface. The term "perforated plasma" refers to a plasma formed during the action of masked heating radiation on the surface of ablating target. In that case, the multi-jet plasma is formed along the target surface.

The energy of the heating pulse decreased after propagation through the MSM. However, the fluence of this radiation on the

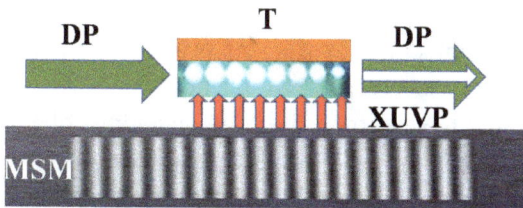

Fig. 4.7. Multi-jet plasma formation using the modulation of the heating pulse. MSM, multi-slit mask; T, target; DP, driving pulse (806 nm); XUVP, extreme ultraviolet pulse. Reproduced from [26] with permission from IOP Publishing.

target surface remained unchanged, since the size of the ablated area was also decreased. It means that the electron and plasma densities in the cases of extended homogeneous and multi-jet plasma were almost equal.

4.2.2. *Analysis of on- and off-axis conditions of QPM*

To analyze the odd and even harmonic generation in both imperforated plasma and multi-jet plasma the TCP scheme was used. The nonlinear optical crystal (BBO, type I, crystal length 0.3 mm, conversion efficiency 5%) was inserted into the vacuum chamber using the translation stage in the path of the focused driving radiation at a distance of 150 mm from the plasma plume. Two-coordinate translating stage allowed the movement of crystal in and out of the path of driving radiation. The driving and second harmonic beams were then focused inside the plasma plume. The polarizations of these two pumps were orthogonal to each other.

These pulses were sufficiently overlapped in the extended plasma to significantly enhance the harmonic spectrum compared with the SCP. The meaning of "enhancement" in that case is related with the well known effect of the influence of TCP on the conversion efficiency of HHG. Briefly, a strong harmonic generation in the case of TCP is possible due to the formation of a quasi-linear field, selection of a short quantum path component, which has a denser electron wave packet, and higher ionization rate compared with the SCP. There is an extensive body of studies dealing with two-color pump in gas phase, which consider many aspects of this driving, including macroscopic effects, polarization-related processes, and commensurate versus incommensurate TCP. Obviously that, for achieving this enhancement, sufficient overlap between two pulses in the plasma area is a necessary requirement.

The estimates have shown that the envelopes of 806 and 403 nm pulses were sufficiently overlapped while entering the extended plasma area. One can assume that the partial overlap decreased the ratio between the interacting second harmonic and driving pulses and diminished, to some extent, the influence of the 403 nm wave on the output spectrum of generating harmonics. However, the influence of second wave was notably strong, since one can easily observe the odd and even harmonics of similar intensity. These studies have shown that the temporal overlap was sufficient for the observation of the peculiarities of the two-color pump using the 806 nm + 403 nm pulses.

Harmonic generation in the extended imperforated plasma using TCP (806 nm + 403 nm) allowed generation of coherent radiation up to 90.75 eV ($\lambda = 13.66$ nm, harmonic cut-off H59, not shown in the upper panel of Fig. 4.8). The definition of harmonic cut-off

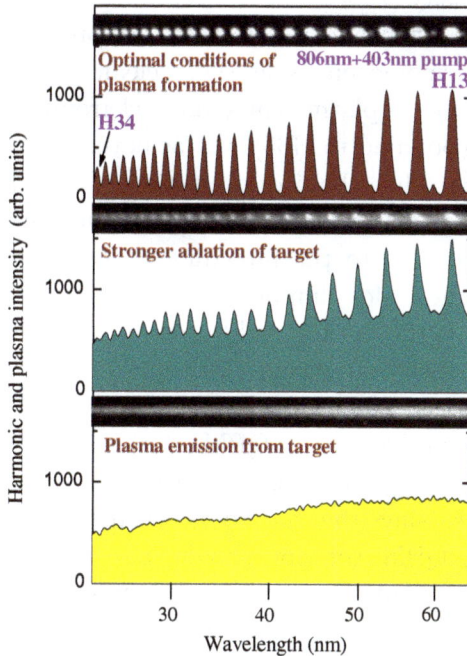

Fig. 4.8. Plasma emission and TCP-induced harmonic spectra generated in the imperforated plasma using the ablation of Ag target by 1.1 J cm^{-2} (upper panel) and 1.3 J cm^{-2} (middle panel) fluencies of heating pulses. Bottom panel shows the plasma emission spectrum observed in the case when the driving pulses did not propagate through the plasma produced at 1.3 J cm^{-2} fluence of heating pulses. Insets to the panels show the raw images of corresponding spectra. The intensity of the 806 nm driving pulse in the focal area was 5×10^{14} W cm^{-2}, while the intensity of the 403 nm pulse in the focal area was estimated to be 2.5×10^{13} W cm^{-2}. Reproduced from [26] with permission from IOP Publishing.

was carried out in the separate experiments. Upper panel of Fig. 4.8 shows the harmonic spectrum at the conditions of target excitation (at fluence $F = 1.1 \, \text{J cm}^{-2}$) when the weak continuum of plasma emission started appearing along with the harmonic emission. The growth of target excitation by stronger fluence of heating pulses ($F = 1.3 \, \text{J cm}^{-2}$) caused the appearance of strong continuum emission followed with the decrease of harmonic intensity and cut-off energy (middle panel). Bottom panel shows the plasma emission observed at similar condition, without the propagation of femtosecond pulses through the plasma medium. Further growth of heating pulse density ($F = 1.8 \, \text{J cm}^{-2}$) stopped HHG while producing solely plasma emission and a few weak low-order harmonics. Best conditions of harmonic emission (i.e. "clean" harmonic spectrum showing the absence of plasma emission background and highest cut-off energy corresponding to H65) was achieved at moderate fluence ($F = 0.9 \, \text{J cm}^{-2}$). Thus the over-excitation of target leading to the growth of plasma density did not lead to the anticipated growth of harmonic yield due to the impeding processes restricting HHG conversion efficiency. Most important of these processes is the phase mismatch between the driving and harmonic waves caused by excess of the electron density of plasma.

As it was shown in previous section, the way to overcome the phase-mismatch problem is a QPM of the waves of driving and harmonic fields by application of perforated plasma. The whole strategy in increasing the harmonic yield in specific ranges of spectrum using QPM concept should be as follows: firstly, the plasma formation should be maintained at "over-excited" conditions, when further increase of heating pulse fluency does not lead to the growth of harmonic yield, and secondly, the division of such plasma formation on a group of separated jets allows the enhancement of a specific group of harmonics at stronger ablation conditions.

Upper panel of Fig. 4.9 shows the spectrum of the harmonics generated at the conditions of the over-excitation of target surface, similarly to the case shown in the middle panel of Fig. 4.8. The spatial modulation of heating beam using different MSMs allowed formation of the multi-jet structures. In the case of the MSM with slit size of

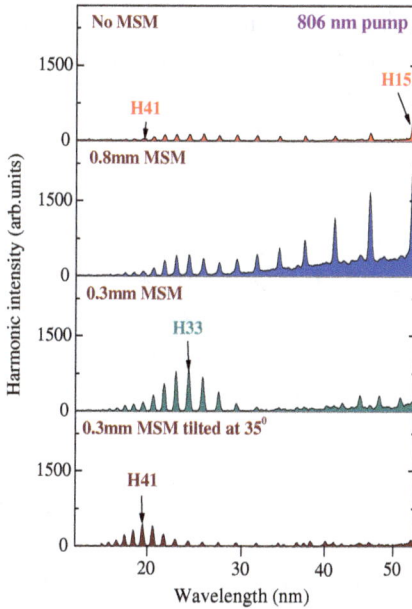

Fig. 4.9. Harmonic spectra in the case of the SCP of extended imperforated plasma (upper panel), three-jet plasma produced using the 0.8 mm MSM (second panel), eight-jet plasma produced using the 0.3 mm MSM (third panel), and ten-jet plasma produced using the 0.3 mm MSM tilted at 35° (fourth panel). The intensity of the driving pulse in the focal area was 5×10^{14} W cm^{-2}. The plasma was produced at the 1.3 J cm^{-2} fluence of heating pulses. Reproduced from [26] with permission from IOP Publishing.

0.8 mm (0.8 mm MSM), the harmonic emission was increased with the highest enhancement occurring for the lower-order harmonics (second panel from the top of Fig. 4.9). Application of smaller sizes of individual jets (0.3 and 0.25 mm) led to the shift of maximally enhanced harmonics towards shorter wavelength region (third and fourth panels). Latter jets were formed by tilting the 0.3 mm MSM at angle $\theta = 35°$. The tilting of MSM at this angle increased the number of jets from 8 to 10 [since the size (l_{jet}) of single jet decreases in accordance with the $l_{jet} = Lcos\theta$, here L is the size of single slit in the used MSM]. A decrease of single jet sizes led to increase of the order of maximally enhanced harmonic (q_{qpm}) assuming the equality of jet sizes and coherence length of this harmonic.

The corresponding maximally enhanced harmonics were H33 and H41. One can expect equal electron densities in the 3-, 8-, and 10-jet plasma configurations assuming the heating pulse has a similar fluence in all these four cases. The electron density in the used conditions of target excitation ($F = 1.3\,\mathrm{J\,cm^{-2}}$) was estimated to be $1.4 \times 10^{17}\,\mathrm{cm^{-3}}$ based on the approach developed in [27]. In accordance with [14], the maximally enhanced harmonics should correspond to H12, H33, and H40 in the case of the multi-jet plasmas with the plasma sizes of 0.8, 0.3, and 0.25 mm respectively. Two calculated values of maximally enhanced harmonics (H33 and H40) agree well with the experimentally observed spectra (see third and fourth panels of Fig. 4.9). As for the case of 0.8-mm jets, the maximally enhanced harmonic (H12) was out of the range of observations in this set of experiments, though one can anticipate this behavior taking into account the growth of enhanced lower-order harmonics in this three-jet plasma (compare the intensities of H15 in two upper panels of Fig. 4.9).

The Y-axes in these experiments were similar for each separate set of experiments (i.e. up to 1900 arbitrary units of intensity in the case of Fig. 4.8, and up to 2800 arbitrary units in the case of Fig. 4.9). The goal of these two figures was to show in one set of line-outs the variation of the harmonic spectra at different conditions of experiments (i.e. with and without the use of MSM).

The harmonic emission was focused inside the XUV spectrometer along the vertical axis using a gold-coated grazing-incidence cylindrical mirror with the glancing angle of $3°$. In that case the raw images of harmonic spectra represented the series of the "dots" corresponding to the distribution of harmonics along the XUV region. Two upper insets of Fig. 4.8 show the examples of such raw images of harmonic spectra captured by CCD camera. While improving the visibility and fluence of generated XUV radiation this method of harmonic images collection did not allow the analysis of the contribution of different parts of driving beam on the spatial distribution of harmonics. To visualize the spatial shapes of harmonics along the vertical axis one has to use the plane mirror instead of the focusing one. Below we present the images of such harmonic spectra and their analysis in the cases of over-excited imperforated and multi-jet plasmas. The aim of these studies was to analyze the variable spatial components of harmonics appearing during frequency conversion in different plasma configurations.

In the case of cylindrical mirror, the spatial, as well as angular distribution of harmonics is masked by the collection of a whole beam in a single spot. In that case, one can hardly distinguish and reveal a difference in the spatio-angular properties of harmonics. The use of flat mirror gives the opportunity in the analysis of those properties. Note that we compared the spatio-angular characteristics of harmonics at variable conditions of plasma excitation. Upper panel of Fig. 4.10 shows the raw image of the H17 to H45 generated in the over-excited imperforated plasma using the above-described method of spectrum collection allowing the observation of the spatio-angular shape of harmonic emission in the on-axis and off-axis regions of driving beam propagation.

The focused driving beam had a Gaussian distribution of intensity. The on-axis part (i.e. central part of focused beam) had stronger intensity. Correspondingly, the conversion of this part of the beam into harmonics showed the difference compared with the off-axis regions of the driving beam, which had lesser intensity. At optimal plasma conditions (i.e. those at which the influence of impeding processes was insignificant) and at intensities up to $\sim 4 \times 10^{14}\,\mathrm{W\,cm^{-2}}$, the harmonics were generated on the on-axis region of the driving beam. With the growth of intensity and/or growth of plasma density (and correspondingly free electron density), the central part of the beam caused the additional ionization of plasma species, which led to further growth of free electron density and phase mismatch. In the meantime, the outer (i.e. off-axis) parts of beam did not cause the growth of the density of the tunnel ionization induced free electrons. For this part of the driving beam, the phase matching conditions remained at the better conditions compared with the on-axis components of the beam, thus allowing harmonic generation.

One can see that the on-axis parts of harmonics were significantly suppressed compared with the off-axis components of the same

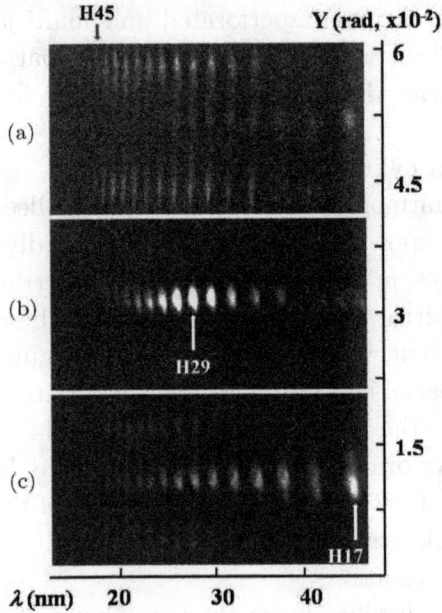

Fig. 4.10. Raw images of SCP-induced HHG spectra in the cases of (a) MSM-free ablation, (b) using MSM with slit sizes of 0.3 mm, and (c) using MSM with slit sizes of 0.8 mm. The vertical axis shows the angular distribution of the images of harmonic spectra obtained on the CCD camera, and the horizontal axis shows the wavelength marks. The intensity of the 806 nm driving pulse in the focal area in the cases of (a-c) were 6×10^{14}, 8×10^{14}, and 8×10^{14} W cm^{-2} respectively. The corresponding plasmas were produced at the (a) 1.3, (b) 1.6, and (c) 1.6 J cm^{-2} fluencies of heating pulses. Reproduced from [26] with permission from IOP Publishing.

orders. Actually, the whole energy of harmonics was concentrated in the off-axis parts of harmonic beams. The ratio of on-axis component with regard to the off-axis one gradually decreased with the growth of harmonic order. The origin of this spatial shape of high-order harmonics is related with the phase mismatching conditions in the axial region of propagation of the driving beam. This phase mismatch increased with the growth of harmonic order. Large electron density led to phase-mismatch and self-defocusing, both of processes leading to deterioration of the optimal conditions of harmonic generation, especially in the central part of beam. Strong intensity of laser radiation led to the appearance of a large amount of tunneled electrons during ionization by the driving pulses. Less intensity of driving beam on the wings of spatial distribution caused weaker influence of the above-mentioned processes on the HHG, which led to the appearance of higher-order harmonics dominantly on the off-axis area. Various aspects of the off-axis driving beam and harmonics have been studied in [28–30].

The installation of the MSM with 0.3-mm slits caused a formation of the perforated plasma. A dramatic re-distribution of harmonics was observed along the spatio-angular distribution of driving radiation (see the raw image presented in the middle panel of Fig. 4.10). The spectrum showed a significant growth of harmonic emission along the on-axis region. The divergence of the enhanced radiation was a few times smaller compared with the case of the HHG in the imperforated plasma. The maximally enhanced harmonic (H29) was significantly stronger compared with the same harmonic order observed in the case of imperforated plasma. In that case, as well as in any other cases of MSM application, the whole plasma length was two times shorter compared with the imperforated plasma, which should lead to a four-fold decrease of harmonic yield instead of the observed QPM-induced enhancement of the group of harmonics. The harmonic emission was concentrated solely on the on-axis area. The weak off-axis components appeared in the harmonic spectrum once we started using the MSM with wider slits (0.8 mm, bottom image of Fig. 4.10), though the on-axis harmonics were significantly stronger compared with the former ones.

Selective QPM of the short and long paths of accelerated electrons responsible for larger and smaller divergence of harmonic emission makes it possible to control their relative weight. The refereed studies of the divergence of QPM-enhanced harmonics from the Ag plasma confirmed this assumption. These harmonics possessed considerably smaller divergence compared with those of lower orders, thus increasing the brightness of shorter-wavelength radiation.

Quasi-phase-matching and quantum-path control of high-order harmonic generation using counterpropagating light has shown the selective QPM of the short and long paths [3]. It was predicted in [31] that the efficiency of generation of soft X-ray harmonics is not necessarily limited by the coherence length derived from the different phase velocities of the driving and harmonic waves. Instead, the biggest contribution to the generated harmonic signal usually comes from those parts of the interaction region where conditions of perfect phase-matching are locally satisfied. These predictions were based on analysis of the paths of the accelerated electron. The assumptions described in [31] had found confirmation in our studies.

To get deeper insight in the spatio-spectral behavior of harmonics, Auguste *et al* [4] have calculated the different contributions of electrons to the phase mismatch. They found that focusing the laser in the nonlinear medium may result in large longitudinal and transverse gradients of the phase that have a major influence on phase matching. Due to the different slopes, the best phase-matching conditions are different for the different trajectories. They have shown that a periodic modulation of the atomic density allows one to quasi-phase-match on-axis the contributions of either the short or the long quantum path. The modulation period, adjusted to twice the coherence length associated with either path, serves as a control parameter. It allows one to dramatically enhance the contribution of the short trajectory to the on-axis emission, through improved beam quality and spectral sharpness.

The intensity distribution of H25-H33 was analyzed in the cases of SCP of plasma shown in upper and middle panels of Fig. 4.10. The vertical line-outs of above harmonics in the case of extended imperforated plasma and eight-jet plasma are shown in Fig. 4.11(a) as the thick and thin curves respectively. The formation of phase-matching conditions along the single jet was again maintained in the following jet leading to the dramatic decrease of divergence and correspondingly the significant growth of the on-axis intensity of harmonics. The divergence of harmonics in that case was six times smaller compared with the case of imperforated plasma. This accumulative effect allowed the emission efficiency of the harmonics surrounding the q_{qpm} to be drastically improved. Figure 4.11(b) shows the integrated spectra of H23-H41 comprising whole harmonic beams in the case of imperforated plasma (upper panel corresponding

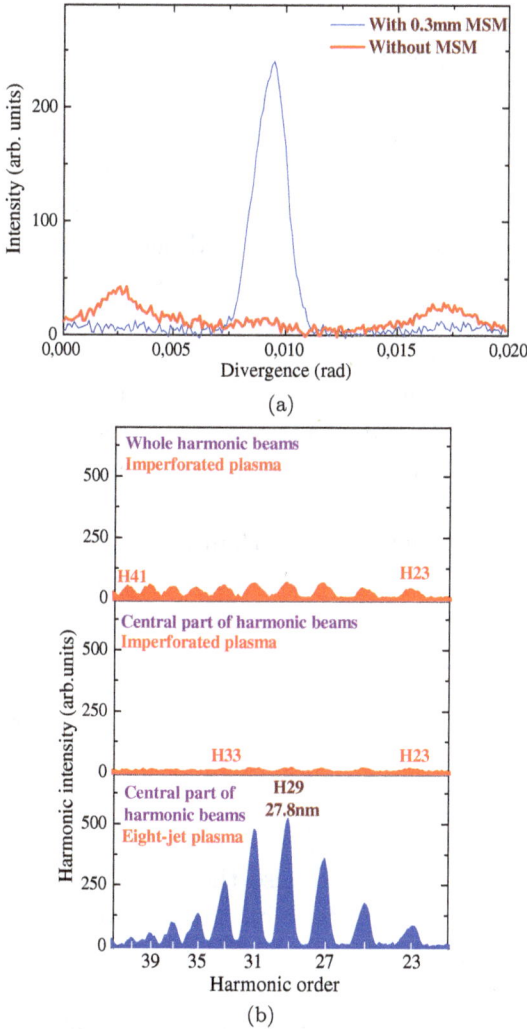

(a)

(b)

Fig. 4.11. (a) The divergences of H25-H33 in the cases of imperforated plasma (thick red curve) and eight-jet plasma. (b) Harmonic spectra comprising (upper panel) whole harmonic beams in the case of imperforated plasma (corresponding to the upper raw image of Fig. 4.10), (middle panel) central part of harmonic beams in the case of imperforated plasma (corresponding to the on-axis emission shown in the upper raw image of Fig. 4.10), and (bottom panel) central part of harmonic beams in the case of eight-jet plasma (corresponding to the middle raw image of Fig. 4.10). Magnification factor of H29 was 25×. The experimental conditions correspond to those presented in the figure caption of Fig. 4.10. Reproduced from [26] with permission from IOP Publishing.

to the upper raw image of Fig. 4.10), central part of harmonic beams in the case of imperforated plasma (middle panel corresponding to the on-axis emission shown in the upper raw image of Fig. 4.10), and central part of harmonic beams in the case of eight-jet plasma (bottom panel corresponding to the middle raw image of Fig. 4.10). The enhancement factor of on-axis H29 in the case of the modulated plasma compared with the imperforated one was 25×.

The goal of these discussed studies was restricted mainly by the analysis of the spatio-angular characteristics of enhanced harmonics at different conditions of QPM. Under optimal conditions of QPM experiments (i.e. at optimal plasma formation and optimal intensity of driving pulses), the enhancement factor was larger than the one reported in this manuscript. The achievement of an enhancement larger than those obtained in present studies was reported in [19, 32], where enhancement factors larger than 30× were achieved.

The QPM in the case of TCP scheme was not as easily achievable as in the case of the SCP. The peculiarities of TCP, compared with SCP, are related with the interaction of the second harmonic wave with the driving fundamental radiation in the plasma plume containing atoms, ions and electrons. The difference of the phases of two waves governs the efficiency of the nonlinear optical processes in this medium. The harmonics in that case are generated mainly from the short trajectories of accelerated electrons. The electrons ionized in this time period are also the main contributors to phase-matched harmonic generation, since the tunneling ionization rate for the two-color field is larger than that for the fundamental field. Consequently, orthogonally polarized two-color field can generate harmonics more efficiently than the fundamental field. However, once we consider the application of the two fields possessing different phases for the formation and maintenance of QPM conditions one has to take into account their separate influence on the optimal phase relations allowing them to overcome the negative influence of propagation effect.

The above-reported QPM-induced enhancement of the group of on-axis harmonics was achieved using the SCP (806 nm). The addition of second field led to a change in the enhancement factor of harmonics. Overall, joint interaction of driving and second harmonic waves inside the imperforated plasma plume allowed significant improvement of harmonic yield compared with the 806 nm pump alone. Figure 4.12(a) shows the growth of harmonic emission in the case of TCP compared with SCP. Three- to four-fold growth of harmonic yield and approximately similar odd and even harmonic generation were among the attractive features of this process in the case of imperforated LPP. However, the application of multi-jet plasma did not allow the QPM of the group of harmonics under these conditions. While SCP easily allowed the conditions of QPM-induced enhancement to be achieved [Fig. 4.12(b), two upper panels], the use of TCP under these conditions did not lead to the same or similar enhancement of the group of harmonics (bottom panel). The difficulties encountered during these studies require additional studies. Below we make some assumptions on that matter, though further steps for improvement of these processes need to be done.

The improvement of QPM under these conditions could be performed using a longer wavelength source of driving radiation, when the phase relations between two pumps become less sensitive for the formation of QPM conditions particularly due to better overlap of two pulses in plasma area and higher second harmonic / driving pulse ratio of pump intensities. To analyze this assumption the 1310 nm pump from optical parametric amplifier and its second harmonic were used for similar experiments using imperforated and ten-jet plasmas. The ratio of assistant field (650 nm) and fundamental field (1310 nm) was 1:4, much larger than in the case of 403 nm + 806 nm pump (1:20). The formation of QPM conditions was achieved in the region of the H39 of 1310 nm pump (see also previous section). The harmonics in the 60–90 nm range were approximately equal to each other. The difference in spectra notably was seen only in the QPM region (30–40 nm).

Note that these spectra were captured at the same collection time, without any changes of the experimental conditions, but just by moving the MSM in or out of the path of heating beam. Thick red curve shows the harmonic spectrum obtained in

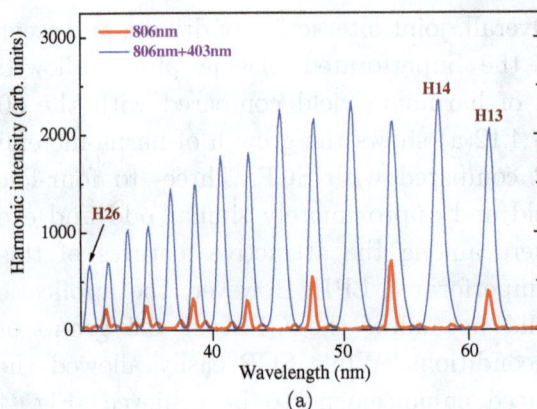

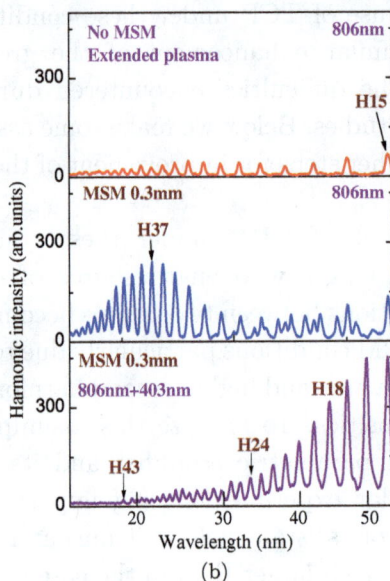

Fig. 4.12. (a) SCP- and TCP-induced spectra generated in the imperforated silver plasma. The intensity of the 806 nm SCP in the focal area was 5×10^{14} W cm^{-2}. The intensities of TCP were 4×10^{14} W cm^{-2} (806 nm) and 2×10^{13} W cm^{-2} (403 nm). (b) Harmonic spectra in the case of the SCP of extended plasma (upper panel; the intensity of the 806 nm pulse in the focal area was 5×10^{14} W cm^{-2}), SCP of eight-jet plasma (middle panel; the intensity of the 806 nm pulse in the focal area was 5×10^{14} W cm^{-2}), and TCP of eight-jet plasma (bottom panel; the intensities of pulses were 4×10^{14} W cm^{-2} (806 nm) and 2×10^{13} W cm^{-2} (403 nm)). Reproduced from [26] with permission from IOP Publishing.

the extended 5-mm-long silver plasma. A featureless spectrum of gradually decreased harmonics starting from the 15th order up to the thirties orders shows the usually observable plateaulike distribution of harmonics. Once the imperforated plasma was modified and formed a group of ten jets by using the MSM placed between the focusing cylindrical lens and target and tilted at $\theta = 35°$, a significant variation of harmonic distribution was observed (thin blue curve). A group of harmonics centered near the H39 was notably enhanced compared with the lower orders. The $\sim 6\times$ enhancement factor was achieved for the maximally enhanced harmonics with regard to the lower-order ones (compare the harmonic intensities in the 35 and 80 nm spectral regions). Simultaneously, the intensity of the QPM-enhanced harmonics was 10 to 20 times stronger compared with the similar ones generated in the imperforated plasma. Probably, the number of harmonics satisfying the conditions of QPM at 1310 nm was larger compared with the case of using 806 nm pump due to larger amount of harmonic orders filling the same spectral region where the QPM was achieved.

In [26], the suggestion about the advantage of using the TCP versus SCP in the case of 1310 nm pump compared with the case of 800 nm pump was analyzed. The use of single MIR photons led to a significant decrease of conversion efficiency due to the λ^{-5} rule. The use of second field dramatically improved the HHG conversion efficiency in the case of MIR + H2 pulses. Note that this improvement was larger than the well-known growth of harmonic yield from 800-nm-class lasers using the TCP (800 nm + H2) in the case of both gases and plasmas.

4.3. Influence of Micro- and Macro-Processes on the High-order Harmonic Generation in Laser-produced Plasma

4.3.1. *Two concepts of MIR-induced harmonic enhancement*

Generation of the high-order harmonics of ultrashort laser pulses by various means (during specular reflection from the surfaces, as well as during propagation through the gases and laser-produced plasmas) has been proven to be an effective method for the formation of coherent short-wavelength sources. The main obstacle in application of these sources is their low fluence, which is caused by small conversion efficiency of the high-order harmonic generation. The history of HHG studies includes various approaches, such as the application of nanoparticles and clusters as the harmonic emitters [33,34], the two-color pump of gases and plasmas [35,36], the application of extended gas-filled fibers [3], the use of feedback loops using various adaptive algorithms [37], the resonance enhancement of single harmonic in the plateau-like region [38], and the quasi-phase-matching of harmonics by different means [8,10,14], which were considered among the most reliable methods for the enhancement of harmonic yield. Currently, the maximal reported conversion efficiencies in gases and plasmas ($10^{-5} - 10^{-4}$) [39,40] have reached the plateau and further growth of this parameter of HHG seems unrealistic. Currently, most of efforts in harmonic generation, at least in the case of gaseous media, were shifted towards the analysis of attosecond pulses generation and their applications, as well as the studies of the molecular orientation of gaseous emitters through the studies of harmonic spectra.

Nevertheless, attempting to grow harmonic yield by different means still remains on agenda. Both abovementioned "old" approaches and "new" proposals in the analysis of harmonic yield enhancement are under consideration in the laboratories dealing with this method of generation of the coherent extreme ultraviolet radiation. The modification of harmonic spectrum may allow various applications in the spectroscopy of harmonic emitters. Analysis of these spectra is crucial for studies of the orientational properties of the molecules existing in the gaseous and plasma-like states [41]. The shift of driving laser pulses towards longer wavelength region offers some perspectives in the extension of harmonic cutoff due to the $E_{\text{cutoff}} \propto \lambda^2$ rule. The analysis of the nonlinear optical response of complex molecules under the action of mid-infrared field allows the specific properties of these species to be defined [42]. Some advances in highly efficient HHG in gases using the MIR pump are shown, for instance, in [43].

The above consideration leads to the advantageous applications of a few methods of harmonic amendment in a single set of experiments. The particular interest here is related with the comparison of the collective mechanisms of accumulation of the nonlinear optical response of medium (so called macro-processes) and the mechanisms related with the individual properties of single emitters (so called micro-processes). Those mechanisms include the fulfillment of the conditions of QPM of the driving and harmonic waves along the whole medium and the conditions of the coincidence of the individual harmonic in the plateau region and the ionic transition possessing strong oscillator strength (gf). The gf value, the product of the oscillator strength f of a transition and the statistical weight g of the lower level, of this transition should be strong enough to cause a significant growth of nearby harmonic. Both these processes could be further amended through the implementation of two-color pump technique in the MIR range.

Though the QPM processes were demonstrated in the gaseous media, no significant resonance-induced enhancement of single harmonic was reported using gas HHG approach. Contrary to that, laser-produced plasmas have proven to be an effective media for these two (macro- and micro-) processes. On the one hand, QPM in LPP has been reported in the case of silver plasma [15], though other ablated materials have also proven to be suitable for these purposes [17]. On the other hand, resonance enhancement of single harmonic has been frequently observed during plasma HHG. In, Sn, Cr, Mn, and a few other, predominantly semiconductor, targets have shown the attractiveness for ablation and further generation of enhanced single harmonic during propagation of the ultrashort pulses through the LPP.

Indium plasma seems a good choice to compare the relative influence of micro- and macro-processes in a single set of experiments. It has demonstrated strong enhancement of single 13th harmonic (H13) of Ti:sapphire laser [44] (see also Chapter 2). The energetic level at 19.92 eV (62.24 nm) corresponding to the $4d^{10}5s^2 {}^1S_0 \rightarrow 4d^9 5s^2 5p(^2D)^1P_1$ transition of In II is exceptionally strong. The gf value of this transition has been calculated to be $gf = 1.11$, which is more than twelve times larger than that of any other transition from the ground state of In II. This transition is energetically close to the 13th harmonic ($h\nu_{13H} = 20.15\,\text{eV}$ or $\lambda = 61.53\,\text{nm}$) of 800 nm radiation, thereby resonantly enhancing its intensity. At the same time, this plasma allows generation of the extended set of harmonics up to the short-wavelength region (H43 of 800 nm radiation, $\lambda = 18.6\,\text{nm}$), which is the requirement for observation of the group of enhanced harmonics due to QPM.

In this section, we demonstrate the advances in using collective processes of enhancement together with the growth of single harmonic induced by the individual properties of emitters for the enhancement of harmonic yield by different means. The indium plasma was used to show the growth of a group of harmonics around the maximally enhanced thirties orders. The tuning of maximally enhanced harmonic was carried out using the variation of the conditions of plasma formation. Simultaneously, we adjust the wavelength of driving pulse to generate the enhanced H21 of the two-color pump (1310 nm–655 nm) using optical parametric amplifier (OPA), which was matched with the abovementioned transition of In II. We present the comparison of these two principally different processes of harmonic enhancement and show the advantages of the joint application of, as much as possible, methods of HHG amendment (QPM-induced enhancement, resonance-induced enhancement, and two-color pump-induced enhancement) in a single experiment [45].

4.3.2. *Comparative enhancement of harmonics caused by resonance enhancement and QPM*

The experimental setup was similar to the one described in section 4.2.1. The majority of the experiments were carried out using the $\sim$1-mJ, 70-fs tunable signal pulses. The signal radiation was used as the driving pulse for the HHG in the LPP. In general experiments were carried out using the two-color pump of LPP. Two pulses were overlapped both temporally and spatially in the extended plasma and allowed a significant enhancement of odd harmonics, as well as the generation of even harmonics with similar intensity as the odd ones. Indium was used as the target for ablation. The size of the target where the ablation occurred was 5 mm. The multi-slit mask was used to create multi-jet plasmas. The size of the slits was 0.3 mm with a distance between them of 0.3 mm.

Below we demonstrate the QPM of the groups of harmonics using the MIR pulses from OPA. Earlier, the principles of the QPM in LPP were demonstrated using the 800 nm lasers. Here we show that similar or even larger enhancement of the group of harmonics generated in indium plasma could be achieved in the case of MIR pump sources, alongside the other methods of HHG amendment.

Tuning of driving radiation allowed the maximal enhancement of single harmonic in the vicinity of 62 nm to be established. Propagation of 1300 nm radiation and its second harmonic through the extended imperforated plasma led to generation of a few strong odd and even harmonics, with maximum enhancement in the case of H21 [Fig. 4.13(a), thick red curve].The spectrum of harmonics is quite unusual, since it contains a group of strong harmonics in the vicinity of the ionic transition of indium possessing large *gf*, contrary to earlier reported generation of single enhanced harmonic using the 800-nm-class lasers [44]. Lower-order harmonics (i.e. those below H21) become weaker with the growth of XUV wavelength, contrary to commonly reported pattern of the plateau-like shape of harmonics demonstrating the gradual decrease of harmonic intensity towards the shorter-wavelength region. The latter behavior was observed in other plasmas used for comparison with the indium plasma. In particular, harmonic generation using two-color pump of silver plasma showed almost plateau-like distribution of harmonics along the broad range of XUV [compare the insets in Fig. 4.13(a) showing the raw images of harmonic spectra in the case of indium (upper panel) and silver (bottom panel) plasmas]. Upper inset of the harmonic spectrum image from indium plasma shown in Fig. 4.13(a) contains the saturated harmonics (particularly H21 and H20). This image is presented in order to clearly underline the difference in the harmonic distribution produced from indium plasma with regard to the plasma generated on the silver target. Notice that all line-outs

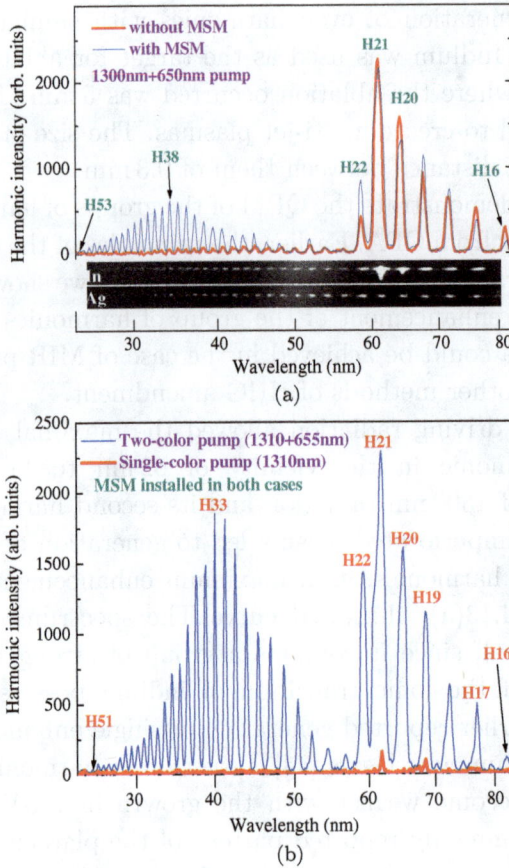

Fig. 4.13. (a) Application of imperforated (thick red curve) and multi-jet (thin blue curve) indium plasmas in the case of two-color (1300 nm + 650 nm) pump. Inset shows two raw images of harmonic spectra obtained in the cases of the two-color pump of expended imperforated indium plasma (upper panel) and silver plasma (bottom panel). (b) Comparative spectra from In multi-jet plasma in the case of single-color (1310 nm, thick red curve) and two-color (1310 nm + 655 nm, thin blue curve) pumps. Reproduced from [45]. Copyright 2014. AIP Publishing LLC.

were taken from the collected spectra where no saturation of registrar was observed. The raw image of the spectrum in the case of indium plasma showed that the enhancement of harmonics occurred at the exact resonance conditions (see maximally enhanced H21) and in the vicinity of this region (H16–H22). Higher orders of harmonics demonstrated almost plateau-like shape starting from the H23 until the cut-off region (H39), similarly to the homogeneous spectrum from silver plasma.

The purpose in showing the raw images in Fig. 4.13(a) is to acquaint the reader with real collection data and visually demonstrate the appearance of the separated group of enhanced harmonics in the case of indium plasma. The spectra were intentionally chosen to be presented as the saturated images for better viewing. The X-axis is shown in the figure on the basis of the calibration of XUV spectrometer. The HHG spectrometer was calibrated using the plasma emission from the used ablated species, as well as other ablating targets.

These two insets showed the harmonic spectra generated using the two-color pump of the homogeneous 5-mm-long indium and silver plasmas produced using the fluence of heating pulse of $F = 1.0\,\mathrm{J\,cm^{-2}}$. Once we separated an extended indium plasma onto a group of eight jets by using the MSM placed in front of ablating indium target, a significant variation of harmonic distribution was observed [Fig. 4.13(a), thin blue curve]. A group of harmonics centered near the H38 was notably enhanced compared with the lower orders. Furthermore, 50× growth of the H38 generated in the multi-jet LPP compared with the same harmonic generated in the extended plasma was achieved. Similar pattern was observed in the case of silver plasma. The analysis of QPM in silver plasma was presented in [15]. Those studies were carried out using the single-color (800 nm) pump. In present studies, the order of QPM-enhanced harmonics became larger compared with the case of using 800 nm pump due to lesser dispersion of plasma in the used MIR region. In the case of MIR pulses (1300 nm), the maximally enhanced harmonic order (q_{qpm}) corresponded to the relation $q_{\mathrm{qpm}} \approx 1.1 \times 10^{18}/(l_{\mathrm{jet}} \times N_{\mathrm{e}})$, where l_{jet} and N_{e} are the length of single plasma jet in multi-jet plume (measured in mm) and electron density of plasma (measured in cm^{-3}) respectively. Remember that the size of each jet was ~0.3 mm.

One can see that lower-order harmonics (H16–H26) in these two cases were approximately similar to each other. Indeed the only difference between these two conditions of plasma formation was the decreased length of the LPP, since the multi-slit mask allowed the ablation of the half of a whole length of target. The saturation of harmonics at large sizes of plasma caused a similar harmonic yield for both 5- and 2.5-mm-long plasmas. The meaning of "saturation" in that case is as follows. The dependence of harmonic yield on the length of extended plasma was analyzed for different harmonic orders. In particular, the $I_{21H}(l_{plasma})$ dependence showed quadratic growth up to $l_{plasma} \approx 1.5\,mm$ and then this dependence had the slope of ~ 0.5 and some instability. The intensities of lower-order harmonics in the case shown in Fig. 4.13(a) are almost similar to each other. However, it is not correct to compare these two spectra from the point of view of variation of harmonic yield as a function of the medium length, though the whole lengths of plasma media used in the cases of homogeneous and perforated plasmas were 5 and 2.5 mm. The small slope of $I_H(l_{plasma})$ for lengths exceeding 1.5 mm (~ 0.5) and additional mechanisms of HHG in the case of multi-jet plasma can affect the yield of lower-order harmonics.

At the used fluence of heating pulse $(1\,J\,cm^{-2})$, the plasma density $(3 \times 10^{17}\,cm^{-3})$ was low enough to exclude the role of absorption processes in the studied spectral region. Thus the saturation, i.e. decrease of the growth of harmonic yield, was attributed to the propagation effect. One can expect a decrease of harmonic yield for larger lengths of medium. Such case could be realized for the on-axis components of harmonic spectra, while off-axis components still became enhanced. Note that our experiments were carried out under the conditions of collection of the whole spectrum of harmonics using the cylindrical mirror inside the XUV spectrometer. However, once the flat mirror was used instead of the cylindrical one, the off- and on-axis distribution of harmonics clearly showed a difference in the harmonic yields for the central and outer parts of the driving beams.

Another situation occurred in the case of higher-order harmonics. While, in the case of imperforated plasma, the intensity of harmonics starting from the 25th order gradually decreased down to the cutoff harmonic ($\sim$H40, thick red curve), the perforated plasma created the preferable conditions for the harmonics centered around the 38th order. The phase matching conditions of the group of harmonics around H38 became significantly improved, which led to notable enhancement compared with the imperforated plasma. The harmonic cutoff was also extended up to H52.

The maximally enhanced harmonic order at the conditions of QPM depended on the fluence of heating pulse. It is obvious that at higher fluence the electron density of plasma became larger and vice versa. Correspondingly, in accordance with above relation, the q_{qpm} at higher fluencies was shifted towards the lower orders. Thin blue curve of Fig. 4.13(b) shows the group of enhanced harmonics centered near H33. This spectrum was obtained at larger fluence compared with the case presented in the upper panel of Fig. 4.13(a) (1.2 and 1 J cm^{-2} respectively). Different wavelengths of MIR pumps in these two cases (1310 and 1300 nm) did not crucially influence the QPM conditions compared with different amount of emitters and free electrons. One can see that QPM-induced enhancement of a group of harmonics became almost equal to the resonance-induced enhancement of single harmonic (compare H33 and H21). Moreover, the whole enhancement of this group of harmonics (between H24 to H48) exceeds the one of harmonics centered around H21.

Once the BBO crystal was removed from the path of MIR radiation the harmonic spectrum from multi-jet plasma became drastically weaker compared with the two-color pump [Fig. 4.13(b), thick red curve]. In the case of single-color pump, the modulation of indium plasma did not improve the harmonic generation efficiency. Only two harmonics were generated (H19 and H21) similarly to the case of extended imperforated plasma, which were far from the spectral range ($\sim$35–45 nm) where the QPM-induced enhancement was expected. Correspondingly, no harmonics were observed in the spectral region where the QPM conditions for the two-color pump were fulfilled. Furthermore, one can see that the resonance enhancement of H21 was significantly decreased in the case of single-color pump. The ratio of enhanced H21 in the case of two-color and single-color pumps was 18. Thus the two-color pump of multi-jet plasma allowed both significant QPM-induced growth of the group

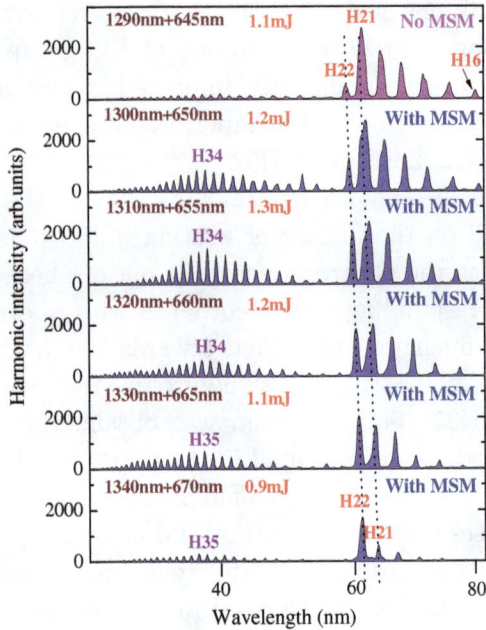

Fig. 4.14. Dependences of the QPM-enhanced harmonics generated in the multi-jet indium plasma on the tuning of the driving MIR pulses. Upper panel shows the harmonic spectrum from imperforated indium plasma. Each panel contains the wavelengths of pumps and the pulse energies. Reproduced from [45]. Copyright 2014. AIP Publishing LLC.

of harmonics and resonance-induced growth of the single harmonic and of a few neighboring orders.

The efficiency of the QPM concept for the plasma harmonics depends on many factors. Figure 4.14 shows the set of harmonic spectra at the conditions of the application of multi-jet indium plasma plumes and tunable MIR radiation. The tuning of MIR pulses along the 1300–1340 nm range led to variation of the intensity of tunable XUV radiation due to the change of the conversion efficiency and amplification for different wavelengths of OPA. Correspondingly, the conversion efficiency of this radiation and its second harmonic towards the odd and even harmonics was changed along the XUV region. The variation of the resonantly-enhanced harmonic generated in multi-jet plasma can be seen by comparing the lower-order harmonics (H16–H22) of the five wavelengths of MIR radiation (1300, 1310, 1320, 1330, and 1340 nm). One can clearly see the redistribution of the relative intensities of the 21st and 22nd harmonics tuned along the strong resonance transition of In II (62.24 nm). In the meantime, the QPM-enhanced harmonics did not tune along the spectral range, while their intensities followed the energies of the MIR pumps. Strongest enhancement of the QPM harmonics was observed in the case of 1310 nm + 655 nm pump (third panel from the top; compare this spectrum with the upper panel obtained in the case of imperforated 5-mm-long plasma pumped by 1290 nm + 645 nm pump). The maximum enhancement was maintained for almost the same order of harmonics (H34 or H35), that points out the independence of the electron density, which defines the q_{qpm}, on the variation of driving pulse intensity. The enhancement of the groups of enhanced harmonics decreased in the case of weaker pump.

Combining two enhancement mechanisms (resonance- and QPM-induced growth of harmonic yield) for the group of harmonic and the single harmonic belonging to this group was tried. In the particular case (indium plasma; MIR pulses; enhanced radiation near 61 nm, which corresponds to approximately H20 or H21 depending on the pump wavelength; separation of the continuous 5-mm-long plasma into eight 0.3-mm-long plasma jets with ∼0.3 mm separation), there is a difference in the wavelengths of resonance-enhanced harmonics (60–65 nm) and QPM-enhanced harmonics (28–45 nm, Fig. 4.13). The first spectral region could not be changed, since it related with the micro-process attributed to the peculiarities of indium ion. The second region is defined by the propagation process described by the relation described in section 4.2 for the maximally enhanced harmonic.

In accordance with this relation, there are two options to move q_{qpm} towards the longer wavelength region. One can either increase the sizes of single jet by increasing the slit sizes of the mask or increase the electron density by increasing the plasma density. Notice that the used intensity of driving pulse ionizes almost every particle within the focal region. We tried the first option by using the MSM containing the 0.8-mm-wide slits. There was the shift of q_{qpm} towards the lower wavelengths. However, the enhancement of the group of harmonics in the case of three 0.8-mm-long jets was not as strong as the one obtained during application of eight 0.3-mm-long plasma jets, probably due to smaller amount of separated emitting media. Notice that the yield of phase matched harmonics quadratically increases with the growth of the number of emitting jets. Second option requires stronger excitation of target and correspondingly larger plasma density. The growth of plasma density causes a larger amount of free electrons to appear through either initial stronger ablation or tunnel ionization of larger amount of plasma particles. However, there is another impeding process, strong incoherent plasma emission in the same XUV region at larger fluence of heating pulse, which decreases the usefulness of generated harmonics.

To realize the dual enhancement of harmonics at the same spectral region, one should use the ions allowing the resonance enhancement of shorter wavelength harmonics. In that case, one can achieve the conditions of mutual appearance of two enhancement processes in the same spectral region, while using large amount of jets. This coincidence of two processes was observed in the tin (see below), chromium, and manganese plasmas, though the enhancement of single harmonics in those plasmas was less pronounced with regard to the indium plasma.

The installation of MSM on the path of heating pulse for multi-jet plasma formation leads to the enhancement of higher-order harmonics only in the case when the phase matching conditions of the harmonic generation in the extended imperforated plasma became worsened due to the presence of large amount of the free electrons strongly affecting the refractive index of plasma. Once the conditions of extended plasma formation were maintained in a proper way (i.e. the plasma contained an insignificant amount of electrons affecting phase matching conditions) the installation of the mask on the path of heating pulse led to a decrease of plasma length followed with the decrease of a whole conversion efficiency.

An example of such a gradual decrease of harmonic yield along the whole spectral range of generation is shown in Fig. 4.15(a). Upper panel shows the harmonic spectrum produced in the 5-mm-long imperforated chromium plasma using the 1310 nm + 655 nm pump. The harmonic generation was obtained at the conditions of Cr target ablation, which did not spoil the phase matching between the interacting waves. Those conditions refer to the availability of further growth of conversion efficiency in the case of longer plasma formation (i.e. at plasma lengths l_{plasma} exceeding 5 mm). The absence of saturation in the $I_H \propto (l_{\text{plasma}})^2$ dependence points out the insignificant influence of various detrimental factors, such as plasma dispersion, which can destroy the coherent accumulation of the generating XUV photons. The modulation of such plasma using MSM, which causes a two-fold decrease of plasma length, leads only to the worsening of HHG conversion conditions due to decrease of the interaction length. This assumption was proven in the case of the modulation of chromium plasma (see bottom panel). The shape of the envelope of harmonic distribution remained the same as in the case of imperforated plasma (upper panel), with the harmonic yield for all orders decreasing by a factor of $\sim$1.5–2.5 (especially for the longer-wavelength part of spectrum).

Another example of the non-optimal use of multi-jet plasma is shown in Fig. 4.15(b). The application of imperforated tin plasma (thick red curve) showed a gradual decrease of harmonic efficiency followed by the enhancement of a group of harmonics in the region of the ionic transition of Sn II possessing large oscillator strength. The enhanced harmonic (H29) almost coincided with the $4d^{10}5s^25p^2P_{3/2} \rightarrow 4d^95s^25p^2(^1D)^2D_{5/2}$ transition of the Sn II ion ($gf = 1.52$ [46]). The studies using the tunable MIR pulses and their second harmonics showed a fine tuning of the resonance-enhanced harmonic and change of the order of this harmonic. The maximally enhanced harmonics, for which both micro- and macro-processes were optimized to generate highest photon yield, were changed from H27 in the case of 1290 nm + 645 nm pump to H31 in the case of 1450 nm + 725 nm pump. In all these cases, the preceding harmonics were suppressed compared with the resonance-enhanced ones [compare H29 and H27 in the case of 1310 nm + 655 nm pump, red curve of Fig. 4.15(b)]. The modulation of extended plasma in that case led to a decrease of lower-order harmonics and the growth of the harmonics

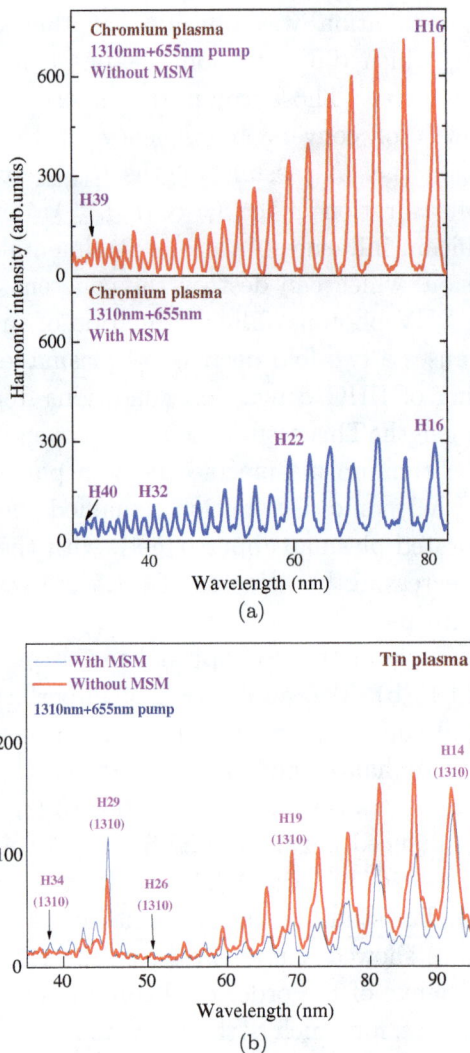

Fig. 4.15. (a) Harmonic spectra generated from the low-ionized chromium plasma using two-color pump. Upper panel was obtained in the case of imperforated 5-mm-long plasma, while bottom panel was obtained in the case of eight-jet plasma. No QPM effect was observed in that case due to small density of free electrons that appeared during laser ablation and tunnel ionization. (b) Harmonic spectra generated in the imperforated (thick red curve) and multi-jet (thin blue curve) tin plasmas. Reproduced from [45]. Copyright 2014. AIP Publishing LLC.

laying around the H30, which was enhanced with a factor of 2. Such small enhancement compared with the case of indium plasma was caused by non-optimal excitation of the target during laser ablation. In other words, once the target was excited and ablated at the conditions when plasma concentration increases together with plasma emission the advantages of the former process became less noticeable compared with the negative influence of strong incoherent plasma emission (from the point of view of further applications of generated coherent radiation).

To justify the usefulness of the harmonics generated using QPM and resonance techniques in indium plasma for various applications one has to present information not only about relative efficiencies when the different parameters change, but also absolute efficiencies of the HHG process and its comparison with coherent XUV radiation generated using other techniques. The comparative studies of gas- and plasma-based HHG have already been carried out in a few laboratories [47–49]. It has been shown that some plasma species, particularly carbon plasma, showed stronger harmonic yield, at least for the lower orders, compared with the argon gas harmonics. One can anticipate that resonantly-enhanced lower-order harmonics from indium plasma would also be strong enough to compete with the gas-based harmonics. The resonantly-enhanced H21 and QPM-enhanced H33 from the In plasma and the harmonics generated in the graphite ablation were compared and found that the former harmonics were as strong as the harmonics generated in the carbon-contained plasma plumes. From this comparative study and previous above-mentioned reports one can conclude about the usefulness in application of these enhanced harmonics in the experiments, similarly to the gas harmonics.

The absolute value of harmonic energy generated from indium plasma was measured by using the following method [50]. Firstly, the third harmonic of Ti:sapphire laser radiation was generated using two nonlinear crystals (BBO, types I and II). The generated third harmonic (266 nm) was measured using an energy meter. Then the third harmonic beam was sent to a vacuum monochromator and converted into visible light using the layer of sodium salycilate deposited on the glass plate. The radiation from this plate was collected by a photomultiplier tube (PMT), and the signal was recorded on an oscilloscope. The signal from PMT was calibrated by the energy of third harmonic measured using an energy meter. Secondly, we compared the third harmonic generated from the plasma using the "monochromator + PMT" registration scheme and the calibrated oscilloscope signal, which allowed us to measure the energy of the third harmonic generated from the plasma. The energy of other high-order harmonics was determined by comparison with the reference third harmonic signal taking into account the spectral response of the monochromator and sodium salycilate. To determine the energy of the higher-order harmonics that were out of the shortest wavelength range of the measurements by this monochromator (80 nm), the above-described homemade XUV spectrometer was used to measure the generated spectrum, thus extrapolating the energy measurements down to the 10 nm region. By using this calibration method, the energy of the QPM-enhanced H33 generated from the indium plasma was measured to be $0.08\,\mu\text{J}$. One can deduce the 8×10^{-5} conversion efficiency of this harmonic taking into account the 1-mJ energy of converted 1310 nm pulses.

4.3.3. *Discussion on different methods of harmonic enhancement*

There are a few issues, which should be addressed for better understanding of the processes occurring during HHG in multi-jet plasmas. In particular, what could cause saturation when absorption by the plasma itself is excluded and quasi-phase matching is centered around H38? Below we address the meaning of saturation with regard to the analyzed processes. There are two options, which may lead to the observation of saturation of the QPM-enhanced harmonics. The first option is related with the limits of coherent adding of the harmonic yield along the whole set of jets. In the experiments, the formation of phase-matching conditions along the single jet was again maintained in the following jet leading to significant growth of the intensity of harmonics. To prove this assumption and the role of QPM in the observed peculiarities of harmonic spectra the intensity (I_H) of harmonics was studied as a function of the number of plasma jets in the case of ablation of target. The heating zones on the target surface were shielded step-by-step to create different number of plasma jets (from single-jet to eight-jet formations).

The anticipated featureless shape of the high-order harmonic spectra from the single 0.3-mm-long plasma jet was similar to those observed in the case of 5-mm-long indium plasma (see insets in Fig. 4.13a and upper panel of Fig. 4.14). With the addition of each next jet, the spectral envelope was drastically changed, with the 38th harmonic intensity in the case of three-jet configuration becoming almost 8 times stronger compared with the case of single-jet plasma. One can anticipate the n^2 growth of harmonic yield for the n-jet configuration compared with the single jet once the phase mismatch becomes suppressed. These conditions allow the growth factor of 9 in the case of three-jet medium to be predicted, which was close to the experimentally measured enhancement factor ($\times 8$). At larger amount of jets, the maximum enhancement factor in the QPM region deviated from the $I_h \propto n^2$ dependence, probably due to unequal properties of the jets, which can arise from the heterogeneous excitation of the extended target. Another reason of the deviation from the quadratic rule could be the growing influence of the driving pulse on the variation of phase relations in the case of a larger amount of contributing zones. In particular, in the case of eight separated zones, we were able to observe the enhancement of harmonics with a factor of $\times 25$, contrary to the anticipated $\times 64$ enhancement. In this connection, the saturation of QPM was caused by abovementioned processes.

The second option is related to the method of spectrum collection. The harmonic emission was focused inside the XUV spectrometer along the vertical axis using the gold-coated cylindrical mirror placed at $87°$ with regard to the interacting radiation. In that case the raw images of harmonic spectra represented the series of the "dots" corresponding to the distribution of harmonics along the XUV region. Two insets of Fig. 4.13a show examples of such raw images of the harmonic spectra captured by CCD camera. While improving the visibility and fluence of generated XUV radiation this method of harmonic images collection did not allow the analysis of the contribution of different parts of driving beam on the spatial distribution of harmonics. To visualize the spatial shapes of harmonics along the vertical axis one has to use the plane mirror instead of the cylindrically focused one. In that case we observed that the on-axis parts of harmonics were significantly suppressed compared with the off-axis components of the same order. The origin of this spatial shape of high-order harmonics is related with the phase mismatch in the axial region of propagation of the driving beam. Strong intensity of laser radiation led to appearance of additional tunneled electrons during ionization by the driving pulses. Less intensity of driving beam on the wings of spatial distribution caused weaker influence of the abovementioned processes on the HHG, which led to appearance of the higher-order harmonics dominantly in the off-axis area. More details regarding on- and off-axis conditions of QPM were described in section 4.2.

The installation of the MSM with 0.3-mm slits caused a dramatic re-distribution of harmonics along the spatial distribution of driving radiation. The spectrum showed a significant growth of harmonic emission along the on-axis region. The divergence of the enhanced radiation was a few times smaller compared with the case of the HHG in the imperforated plasma. The maximally enhanced harmonic (H38) in the on-axis area was significantly stronger compared with the same harmonic order observed in the case of imperforated plasma. The harmonic emission was concentrated solely on the on-axis area. The meaning of the saturation of QPM in that case was related with the re-distribution of this process along the spatial shape of converting beam. The conditions of QPM were far from optimal for different parts of driving beam.

The next issue is related with the necessity in formation of appropriate conditions for the HHG and QPM. Figure 4.15(b) shows the decreased low-order harmonics from tin plasma for the spectrum with the multi-slit mask, as expected for shorter interaction lengths below the coherence length. In contrast there is an increased signal for a few harmonics around the resonance. This result points out the relation of the decrease of lower-order harmonics in the case of using the MSM with the two-fold decrease of the whole length of plasma. Meanwhile the higher-order harmonics (between H28 to H34) are enhanced due to formation of the quasi-phase matching conditions. However, there are different ranges of these conditions. Under ideal conditions, when the coherence lengths of H29 or H30 exactly match with the sizes of jets, one can expect a significant gain of those harmonics. However, once the harmonic yield does not coherently add along the whole set of jets, partially due to imperfect distribution of atoms/ions and free electrons in each jet, the QPM conditions became deteriorated to some extent. That is why the observed enhancement factor was only ×2. However, even under these "non-optimal" conditions of plasma formation we were able to observe the multi-particle-induced growth of the harmonic, which was notably suppressed by the single-particle-induced process (compare H28 in these two cases).

4.4. Conclusions to Chapter 4

We have shown the quasi-phase-matching of a group of harmonics generated in the multi-jet plasma produced by laser ablation of bulk silver target using tunable pulses in the region of 1250–1400 nm and their second harmonic emission. We have discussed the observation of 12-fold growth of shorter-wavelength harmonics compared with longer wavelength ones in the eight-jet plasma and more than 60× growth of 39th harmonic in multi-jet LPP compared with extended plasma. We have also shown the numerical treatment of this effect, which included microscopic description of the harmonic generation, propagation of the pump pulse, and propagation of the generated harmonics. This theoretical treatment has shown qualitative agreement with the observed peculiarities of modulated harmonic spectra from the multi-jet plasma.

Further, we have analyzed the spatial features of the QPM in LPP at over-excitation of silver plasma using the 806 nm driving pulses and second harmonic. The studies of the off-axis and on-axis components of harmonics allowed the demonstration of significantly enhanced harmonics in the latter case. The divergence of enhanced radiation was a few times smaller compared with the case of the HHG in imperforated plasma. The enhancement factor of on-axis 29th harmonic in the case of modulated plasma was 25× compared with the imperforated plasma.

The TCP-induced HHG at the conditions of multi-jet plasma have shown the complexity of the phase relations between two orthogonally polarized pumps and harmonic waves. The comparative studies of this process using 806 and 1310 nm pulses and their second harmonics have shown better conditions for the formation of QPM in the case of latter pump, which could be explained by weaker influence of relative phase between the interacting waves in the case of longer-wavelength source, better overlap of these waves in the plasma area, and larger ratio of assistant field compared with the fundamental one.

Finally, indium plasma has proven to be a suitable medium to compare the enhancement of harmonics caused by micro-processes (i.e. resonance-induced enhancement of a group of odd and even harmonics) and macro-processes (i.e. quasi-phase-matching induced growth of the group of harmonics using the separated plasma jets) under the conditions of the MIR-induced two-color pump of modulated medium. At the same time, attention should be taken for the proper formation of the plasma suitable for demonstration of the QPM. It was shown that no QPM conditions were achieved at the small plasma and electron density of Cr ablation.

Forming resonance conditions to enhance the nonlinear optical response of the medium may be an alternative to the phase-matching technique previously used for harmonic enhancement. In the meantime, if this process could be combined with the phase-matching effects and/or coherent control of HHG, one will be able to generate the harmonics enhanced by both collective and individual features of the medium. Application of tunable MIR pulses is the additional advantage in this direction since, on the one hand, the tuning of driving pulses towards the wavelength the integers of which exactly coincide with the ionic transitions possessing large *gf* allows the maximal resonantly-enhanced yield of single harmonic to be achieved. On the other hand, the use of longer-wavelength laser sources allows the extension of harmonic cutoff and correspondingly the observation of QPM effect in the shorter-wavelength region. Indium plasma perfectly matched with above conditions and became the best choice for comparative studies to answer the question of which (either collective or individual) process allows stronger nonlinear optical response of the medium. Further amendments are achieved once one uses the two-color pump technique for these two mechanisms.

References

[1] J. A. Armstrong, N. Bloembergen, J. Ducuing, and P. S. J. Pershan, *Phys. Rev.* **127**, 1918 (1962).

[2] A. Paul, R. A. Bartels, R. Tobey, H. Green, S. Weiman, I. P. Christov, M. M. Murnane, H. C. Kapteyn, and S. Backus, *Nature* **421**, 51 (2003).

[3] X. Zhang, A. L. Lytle, T.Popmintchev, X. Zhou, H. C. Kaptayn, M. M. Murnane, and O. Cohen, *Nature Phys.* **3**, 270 (2007).

[4] T. Auguste, B. Carré, and P. Salières, *Phys. Rev. A* **76**, 011802 (2007).

[5] J. Seres, V. S. Yakovlev, E. Seres, C. H. Streli, P. Wobrauschek, C. H. Spielmann, and F. Krausz, *Nature Phys.* **3**, 878 (2007).

[6] M. Zepf, B. Dromey, M. Landreman, P. Foster, and S. M. Hooker, *Phys. Rev. Lett.* **99**, 143901 (2007).

[7] A. L. Lytle, X. Zhang, R. L. Sandberg, O. Cohen, H. C. Kapteyn, and M. M. Murnane, *Opt. Express* **16**, 6544 (2008).

[8] A.Pirri, C. Corsi, M. and Bellini, *Phys. Rev. A* **78**, 011801 (2008).

[9] V. Tosa, V. S. Yakovlev, and F. Krausz, *New J. Phys.* **10**, 025016 (2008).

[10] A.Bahabad, M. M. Murnane, and H. C. Kapteyn, *Nature Phys.* **4**, 570 (2010).

[11] A. Willner, F. Tavella, M. Yeung, T. Dzelzainis, C. Kamperidis, M. Bakarezos, D. Adams, M. Schulz, R. Riedel, M. C. Hoffmann, W. Hu, J. Rossbach, M. Drescher, N. A. Papadogiannis, M. Tatarakis, B. Dromey, and M. Zepf, *Phys. Rev. Lett.* **107**, 175002 (2011).

[12] K. O'Keeffe, T. Robinson, and S. M. Hooker, *Opt. Express* **20**, 6236 (2012).

[13] T. Fok, Ł. Wégrzyński, M. Kozlova, J. Nejdl, P. W. Wachulak, R. Jarocki, A. Bartnik, and H. Fiedorovicz, *Photonics Lett. Pol.* **6**, 14 (2014).

[14] R. A. Ganeev, M. Suzuki, and H. Kuroda, *Phys. Rev. A* **89**, 033821 (2014).

[15] R. A. Ganeev, V. Tosa, K. Kovács, M. Suzuki, S. Yoneya, and H. Kuroda, *Phys. Rev. A* **91**, 043823 (2015).

[16] A.Husakou and J. Herrmann, *Phys. Rev. A* **90**, 023831 (2014).

[17] R. A. Ganeev, M. Suzuki, and H. Kuroda, *J. Phys. B: At. Mol. Opt. Phys.* **47**, 105401 (2014).

[18] R. A. Ganeev, M. Suzuki, and H. Kuroda, *Eur. Phys. J. D* **70**, 21 (2016).

[19] R. A. Ganeev, A. Husakou, M. Suzuki, and H. Kuroda, *Opt. Express* **24**, 3414 (2016).

[20] C. Kan, N. H. Burnett, C. E. Capjack, and R. Rankin, *Phys. Rev. Lett.* **79**, 2971 (1997).

[21] A. V. Andreev, R. A. Ganeev, H. Kuroda, S. Y. Stremoukhov, and O. A. Shoutova, *Eur. Phys. J. D* **67**, 22 (2013).

[22] L. Zheng, X. Chen, S. Tang, and R. Li, *Opt. Express* **15**, 17985–17990 (2007).

[23] M. Lewenstein, Ph. Balcou, M. Y. Ivanov, Anne L'Huillier, and P. B. Corkum, *Phys. Rev. A* **49**, 2117 (1994).

[24] I.J. Kim, C. M. Kim, H. T. Kim, G. H. Lee, Y. S. Lee, J. Y. Park, D. J. Cho, and C. H. Nam, *Phys. Rev. Lett.* **94**, 243901 (2005).

[25] J. Mauritsson, P. Johnsson, E. Gustafsson, A. L'Huillier, K. J. Schafer, and M. B. Gaarde, *Phys. Rev. Lett.* **97**, 013001 (2006).

[26] R. A. Ganeev, *J. Phys. B: At. Mol. Opt. Phys.* **49**, 095402 (2016).

[27] R. A. Ganeev, M. Suzuki, S. Yoneya, and H. Kuroda, *Laser Phys.* **24**, 115405 (2014).

[28] E. Takahashi, Y. Nabekawa, and K. Midorikawa, *Opt. Lett.* **27**, 1920 (2002).

[29] L. V. Dao, S. Teichmann, J. Davis, and P. Hannaford, *J. Appl. Phys.* **104**, 023105 (2008).

[30] S. Teichmann, P. Hannaford, and L. V. Dao, *Appl. Phys. Lett.* **94**, 171111 (2009).

[31] V. S. Yakovlev, M. Ivanov, and F. Krausz, *Opt. Express* **15**, 15351 (2007).

[32] R. A. Ganeev, M. Suzuki, P. V. Redkin, and H. Kuroda, *J. Nonlin. Opt. Phys. Mater.* **23**, 450013 (2014).

[33] T. D. Donnelly, T. Ditmire, K. Neuman, M. D. Perry, and R. W. Falcone, *Phys. Rev. Lett.* **76**, 2472 (1996).

[34] R. A. Ganeev, M. Baba, M. Suzuki, and H. Kuroda, *J. Phys. B* **47**, 135401 (2014).

[35] J. Kim, C. M. Kim, H. T. Kim, G. H. Lee, Y. S. Lee, J. Y. Park, D. J. Cho, and C. H. Nam, *Phys. Rev. Lett.* **94**, 243901 (2005).

[36] R. A. Ganeev, H. Singhal, P. A. Naik, J. A. Chakera, H. S. Vora, R. A. Khan, and P. D. Gupta, *Phys. Rev. A* **82**, 053831 (2010).

[37] C. Winterfeldt, C. Spielmann, and G. Gerber, *Rev. Mod. Phys.* **80**, 117 (2008).

[38] R. A. Ganeev, V. V. Strelkov, C. Hutchison, A. Zaïr, D. Kilbane, M. A. Khokhlova and J. P. Marangos, *Phys. Rev. A* **85**, 023832 (2012).

[39] E. Constant, D. Garzella, P. Breger, E. Mével, Ch. Dorrer, C. Le Blanc, F. Salin, and P. Agostini, *Phys. Rev. Lett.* **82**, 1668 (1999).

[40] L. B. Elouga Bom, J.-C. Kieffer, R. A. Ganeev, M. Suzuki, H. Kuroda, and T. Ozaki, *Phys. Rev. A* **75**, 033804 (2007).

[41] C. B. Madsen and L. B. Madsen, *Phys. Rev. A* **74**, 023403 (2006).

[42] C. Hutchison, R. A. Ganeev, M. Castillejo, I. Lopez-Quintas, A. Zair, S. J. Weber, F. McGrath, Z. Abdelrahman, M. Oppermann, M Martín, D. Y. Lei, S. A. Maier, J. W. Tisch, and J. P. Marangos, *Phys. Chem. Chem. Phys.* **15**, 12308 (2013).

[43] T. Popmintchev, M.-C. Chen, D. Popmintchev, P. Arpin, S. Brown, S. Ališauskas, G. Andriukaitis, T. Balčiunas, O. D Mücke, A. Pugzlys, A. Baltuška, B. Shim, S. E Schrauth, A. Gaeta, C. Hernández-García, L. Plaja, A. Becker, A. Jaron-Becker, M. M. Murnane and H. C. Kapteyn, *Science* **336**, 1287 (2012).

[44] R. A. Ganeev, M. Suzuki, M. Baba, H. Kuroda and T. Ozaki, *Opt. Lett.* **31**, 1699 (2006).

[45] R. A. Ganeev, *J. Appl. Phys.* **119**, 113104 (2016).

[46] G. Duffy, P. van Kampen, and P. Dunne, *J. Phys. B* **34**, 3171 (2001).

[47] R. A. Ganeev, T. Witting, C. Hutchison, F. Frank, P. V. Redkin, W. A. Okell, D. Y. Lei, T. Roschuk, S. A. Maier, J. P. Marangos and J. W. G. Tisch, *Phys. Rev. A* **85**, 015807 (2012).

[48] Y. Pertot, S. Chen, S. D. Khan, L. B. Elouga Bom, T. Ozaki and Z. Chang, *J. Phys. B: At. Mol. Opt. Phys.* **45**, 074017 (2012).

[49] M. Wöstmann, P. V. Redkin, J. Zheng, H. Witte, R. A. Ganeev and H. Zacharias, *Appl. Phys. B* **120**, 17 (2015).

[50] R. A. Ganeev, M. Baba, M. Suzuki and H. Kuroda, *Phys. Lett. A* **339**, 103 (2005).

Chapter 5

Various Applications of MIR HHG Approach in Different Plasma Plumes

In this Chapter, we show different schemes and applications of MIR HHG in plasmas and compare them with the use of fixed driving sources. In particular, we demonstrate the generation of harmonics up to the 27th order ($\lambda = 29.9\,\text{nm}$) of 806 nm radiation in boron carbide plasma. We analyze the advantages and disadvantages of this target compared with the ingredients comprising B_4C (solid boron and graphite) by studying the plasma emission and harmonic spectra from three species. We compare different schemes of the two-color pump of B_4C plasma, particularly using the second harmonics of 806 nm laser and optical parametric amplifier (1310 nm) as the assistant fields, as well as demonstrate the sum and difference frequency generation using the mixture of the wavelengths of two laser sources. These studies showed the advantages of the two-color pump of B_4C plasma leading to stable harmonic generation and growth of harmonic conversion efficiency. We also show that the coincidence of harmonic and plasma emission wavelengths in most cases does not cause the enhancement or decrease of the conversion efficiency of this harmonic. The spatial characterization of harmonics shows their on-axis modification depending on the conditions of frequency conversion.

Further we discuss the high-order harmonic generation in silver, gold, and zinc plasma plumes irradiated by orthogonally polarized two-color MIR field. We analyze an increase of the HHG efficiency in comparison with the single-color case, which essentially depends on the plasma species and harmonic order. An increase of more than an order of magnitude is observed for silver plasma, whereas for gold and zinc it is lower; these results are reproduced in the calculations that include both propagation and microscopic response studies. We show that the widely used theoretical approach assuming the $1s$ ground state of the generating particle fails to reproduce the experimental results; agreement is achieved in the theory using the actual quantum numbers of the outer electron of the generating particles. Moreover, the theoretical studies highlight the redistribution of the electronic density in the continuum wave packet as an important aspect of the HHG enhancement in the two-color orthogonally polarized fields with comparable intensities: in the single-color field the electronic trajectories with almost zero return energy are the most populated ones; in the two-color case the total field maximum can be shifted in time so that the trajectories with high return energies (in particular, the cut-off trajectory) become the most populated ones.

Finally, we demonstrate high-order sum and difference harmonic generation using the mixing of tunable (1200–1600 nm) pulses from optical parametric amplifier and 810 nm ultrashort pulses in the extended laser-produced graphite plasma. Optimization of high-order harmonic generation using various parameters (delay between two-color incommensurate and commensurate waves, ratio between pulse intensities, use of parallel and orthogonal polarizations of interacting pulses, etc.) and comparison with two-color pump using mid-infrared and second-harmonic pulses are presented. We show that, despite non-optimal spatio-temporal overlap of two or three sources of radiation in the plasmas, the application of proposed technique allows generation of significantly denser combinations of interacting waves leading to sum and difference frequencies generation in the extreme ultraviolet region.

5.1. Ablation of Boron Carbide for High-order Harmonic Generation of Ultrafast Pulses in Laser-produced Plasma

5.1.1. *Application of hard materials for plasma formation*

Boron carbide (B_4C) is an extremely hard chemical material used in tank armor, bullet-proof vests, and numerous industrial applications, such as chemical deposition, laser-assisted deposition, and magnetron sputtering. With a hardness of 9.3 on the Moh's scale, it is one of the hardest materials known, behind cubic boron nitride and diamond. As underlined in [1], pulsed laser deposition of boron carbide has been explored only to a limited extent, and may be of high value since laser ablation easily overcomes the difficulties associated with the high melting point of B_4C and allows for a broad range of deposition conditions through the control of the laser and plasma parameters. Analysis of the optical and nonlinear optical properties of the laser-produced plasma formed on the surface of bulk B_4C can give some insight in the usefulness of this medium for different applications. Particularly, the studies presented in [1–4] have shown that the low-order harmonic generation in boron carbide plasma is sensitive to the presence of atoms, molecules, clusters and nanoparticles in a LPP, and can, in some cases, be used as a probe of their density.

The harmonic generation in boron carbide LPP was not limited only to generation of lower-order harmonics. Recent reports showed the advantages in using this plasma for generation of coherent emission in the shorter wavelength region. In particular, this LPP could be used as the effective medium for high-order harmonic generation. First attempts to analyze the high-order nonlinear optical properties of the LPP produced on the surface of boron carbide were demonstrated in [5,6]. Those studies have shown the attractive properties of B_4C plasma as the medium for HHG. The important peculiarity of the ablation of this material is the insignificant modification of target surface due to high temperature of melting and hardness of the surface, which allows the stability of plasma and harmonic characteristics to be maintained. Earlier studies have shown that LPP can be used as a nonlinear medium for the HHG if the effects of the limiting factors (i.e. self-phase modulation, self-defocusing, and the phase mismatch induced by the abundance of free electrons in the plasma plumes) are minimized [7]. There are plenty of the specific features of the HHG in plasma plumes, the foremost one is the wide range of nonlinear medium characteristics available for varying the conditions for the formation of the laser plume on the surface of solid targets.

A search of new plasma media and the definition of the best methods of extended plasma formation are the options for further enhancement of harmonic yield in the extreme ultraviolet range. A substantial increase in the highest order of generated harmonics, the emergence of a second plateau in the intensity distribution of harmonics, the high efficiencies obtained with several plasma plumes, the realization of resonance enhancement of individual harmonics, the formation of quasi-phase-matching conditions in the extended plasma plume, the efficient harmonic enhancement in the plasmas containing the clusters of different materials, and other advantages demonstrated in previous studies of the harmonics generated in the LPP make it reasonable to further analyze the specially prepared plasmas for HHG. In this connection, the molecular targets, such as B_4C, may be of interest in comparison with the atomic ingredients of their compounds.

In this section, we discuss the reported demonstration of the harmonic generation up to the 27th order (H27, $\lambda = 29.9$ nm) of 806 nm driving pulses in the boron carbide LPP. We analyze the advantages and disadvantages of this target compared with the atomic ingredients comprising B_4C. We compare different schemes of the two-color pump of B_4C plasma. We also show that the coincidence of harmonic and plasma emission wavelengths in most cases does not lead to enhancement or decrease of the conversion efficiency of this harmonic. The characterization of the spatial properties of harmonics shows that their modification depends on

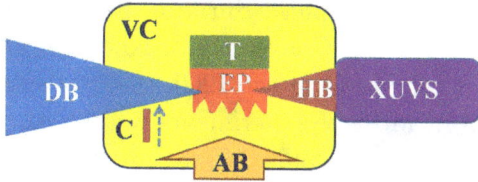

Fig. 5.1. Experimental setup for harmonic generation in B_4C plasma. DB, driving beam; AB, ablation beam; VC, vacuum chamber; T, target; EP, extended plasma; HB, harmonic beam; C, BBO crystal; XUVS, extreme ultraviolet spectrometer. Reproduced from [8] with permission from Elsevier.

the conditions of HHG, which allowed the definition of the impeding processes, which diminish the HHG conversion efficiency [8].

5.1.2. *Experimental arrangements*

Two configurations of laser sources. In the first case, the uncompressed radiation of Ti:sapphire laser (central wavelength 806 nm, pulse duration 370 ps, pulse energy up to $E_{hp} = 6$ mJ, 10 Hz pulse repetition rate) was used for the extended plasma formation. These heating pulses were focused using a 200 mm focal length cylindrical lens inside the vacuum chamber containing the ablating slab targets to create the extended plasma plume (Fig. 5.1). The intensity of the heating pulses on the target surface was varied up to $I_{hp} = 4 \times 10^9$ W cm^{-2}. The plasma sizes were 5×0.08 mm^2. The compressed driving pulses from the same laser with energy up to $E_{dp} = 3$ mJ and 64 fs pulse duration were focused inside this extended plasma for harmonic generation. The intensity of the driving pulses at the focus area was varied up to 4×10^{14} W cm^{-2}.

In the second case, the experimental setup comprised the Ti:sapphire laser, traveling-wave optical parametric amplifier (OPA), and HHG scheme using propagation of two pulses (amplified signal or idler radiation from OPA and 806 nm radiation) through the extended LPP, similarly to the experiments analyzed in Chapters 3 and 4. Part of the amplified uncompressed radiation of the Ti:sapphire pump laser (pulse energy of 5 mJ) was separated from a whole beam and used as a heating pulse for homogeneous extended plasma formation. The compressed radiation of Ti:sapphire laser (pulse energy 8 mJ, pulse duration 64 fs) pumped the OPA (HE-TOPAS Prime, Light Conversion). Signal and idler pulses from OPA allowed the tuning along the 1200–1600 nm and 1600–2600 nm ranges respectively. Most of experiments were carried out using the mixture of 1 mJ, 70 fs signal pulses (1310 nm) and 3 mJ, 64 fs, 810 nm pulses in the boron carbide plasma. The intensity of focused 1310 nm pulses inside the LPP was 2×10^{14} W cm^{-2}.

The length of interaction zone was 5 mm. The confocal parameter was 12 mm, so the experiments were carried out under the conditions of loosely focused femtosecond radiation. The optimal distance between the target and focal spot of the main beam is directly related to the time delay between the heating and driving pulses. The optimal conditions for harmonic generation in boron carbide plasma at 30 ns delay were found to be at the distance of ~150 μm from the target surface.

The plasma and harmonic emissions were analyzed using the XUV spectrometer containing the replaceable cylindrical or plane mirrors and a 1200 grooves/mm flat field grating with variable line spacing. The spectrum was recorded on a micro-channel plate detector with the phosphor screen, which was imaged onto a CCD camera.

The harmonic generation of the laser pulses propagating through the plasma produced during ablation of boron carbide was studied. We also compared the plasma and harmonic emission from this ablation with the harmonics generated in the plasmas produced on the boron and graphite targets. The sizes of all targets were 5 mm. The B$_4$C and other slab targets (purity 99.9%) were mounted on a three-coordinate holder placed inside a vacuum chamber (~10^{-5} mbar) to adjust with regard to the propagating femto-second pulse.

5.1.3. *Comparison of the plasma and harmonic emission of ablated targets*

Plasma emission during over-excitation of B$_4$C target comprised the lines attributed to the B II-B IV and C II-C III ions

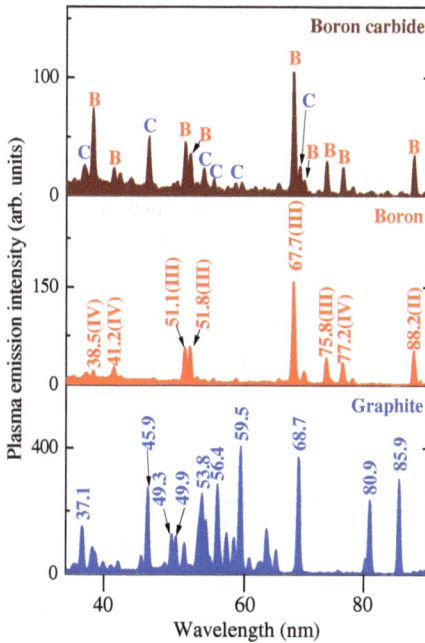

Fig. 5.2. Plasma spectra from boron carbide (upper panel), boron (middle panel), and graphite (bottom panel) ablations at similar fluence of heating pulses ($F = 1.7\,\mathrm{J\,cm^{-2}}$). These plasma spectra were time integrated. One can see that the former target showed the weakest plasma emission (compare Y-axes of three panels). These conditions of plasma formation did not allow the efficient generation of harmonics due to large amount of free electrons in the plasma plumes. Reproduced from [8] with permission from Elsevier.

(Fig. 5.2, upper panel). These conditions of target ablation (fluence $F = 1.7\,\mathrm{J\,cm^{-2}}$) did not allow efficient harmonic generation in the LPP, since large amount of free electrons led to phase mismatch between the interacting waves during HHG. Similar excitation of boron and graphite targets showed the emission lines (Fig. 5.2, middle and bottom panels), which were identical to those observed during ablation of boron carbide. The latter target showed the weakest plasma emission with regard to the B and C targets (compare Y-axes of three panels) at equal conditions of ablation. This difference in plasma emission was caused by higher melting point of the boron carbide compared with the boron and graphite targets. Note that these conditions of plasma formation led to extremely weak harmonic generation in the B and C plasmas as well.

Optimal fluence for ablation of these three targets was chosen by achieving strongest harmonic yield. Graphite allowed generation of the lower-order harmonics with largest conversion efficiency. At the same time, the boron plasma, while showing the lowest conversion efficiency among the used species, generated harmonics with highest cut-off (H59 of 806 nm pump, $\lambda = 13.7$ nm). Other two plasmas showed significantly lower harmonic cut-offs (H27 and H25 in the case of B_4C and C plasmas respectively). The comparison of these plasma formations, from the point of view of the stability of harmonic emission, showed that boron carbide plasma was the best choice among three ablated targets.

These observations demonstrated that, in spite of a similarity of plasma emission characteristics in the case of these two pairs (i.e. B_4C-B and B_4C-C, Fig. 5.2), the frequency conversion in boron carbide showed a difference with regard to other plasmas. The most probable reason in the difference of harmonic emission properties of B_4C and two other ablated targets is the difference in the constituencies of plasmas in these two pairs under the optimal conditions of laser ablation. Probably, the neutral molecules of boron carbide play an important role in harmonic generation. Their involvement in HHG may explain the low harmonic cut-off observed in the case of B_4C plasma. Similarly, the lower conversion efficiency in B_4C plasma, compared with graphite plasma, could be explained by preferable involvement of neutral molecules in harmonic generation, while in the case of graphite ablation one can expect the appearance of the carbon clusters, which enhance the harmonic emission [9–11].

The observed difference in the HHG from the plasmas produced on the boron carbide and its ingredients can also be attributed to different influence of the neutral and ionic species (B, C), compared with the B_4C molecules, on the process of frequency conversion. It was earlier shown in [7] that the seventh harmonic of Nd:YAG laser ($\lambda = 152$ nm) generating in the B_4C plasma did not demonstrate the enhancement compared with other carbon-containing plasma species showing the four to five fold growth of this harmonic compared with the lower ones. The study reported in [8] demonstrated that the difference in the nonlinear optical properties of the plasmas formed on the surfaces of boron carbide and its ingredients (B and C) points to the presence of B_4C molecules in the plasma plume without the disintegration during laser ablation. Those and present observations show that plasma HHG can be considered as a tool for the nonlinear spectroscopy.

5.1.4. *Double-pulse and two-color pump of boron carbide plasma*

In [8], different techniques of the HHG in the B_4C plasma were analyzed. One of them (double-pulse technique) has recently been introduced for the studies of plasma harmonics [12]. This technique allows the HHG using the two femtosecond pulses propagated in the vicinity of solid targets. Below we briefly describe the principles of this method allowing the characterization of the targets without the preliminary ablation using the heating picosecond pulses. Two pulses from the same Ti:sapphire laser were obtained by tuning the trigger signal on the Pockels cell. Commonly, this cell was used for separating the single laser pulse from the train of amplified pulses generated in the regenerative amplifier. By varying the trigger signal one can separate two pulses. Different ratios of the intensities of the first and second pulses separated by 8 ns from each other were obtained in these experiments. The optimal ratio in the case of boron carbide was 0.7. First pulse propagated in the vicinity of target and ablated it by the spatial wing of the intensity distribution in the focal area. Then second pulse propagated through the plasma, which led to harmonic generation. Figure 5.3(a) (thick curve) shows a few harmonics (H13–H23) generated in that case. The harmonic efficiency and harmonic cut-off using the double-pulse technique were predictably lower compared with the conventional method of optimal plasma formation followed by the propagation of the optimally delayed femtosecond pulse through the LPP. In the meantime, the simplicity, the absence of the additional source of heating radiation,

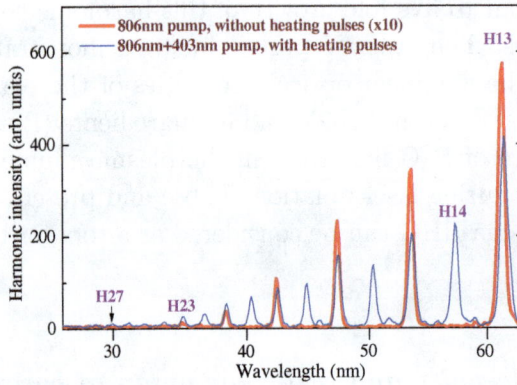

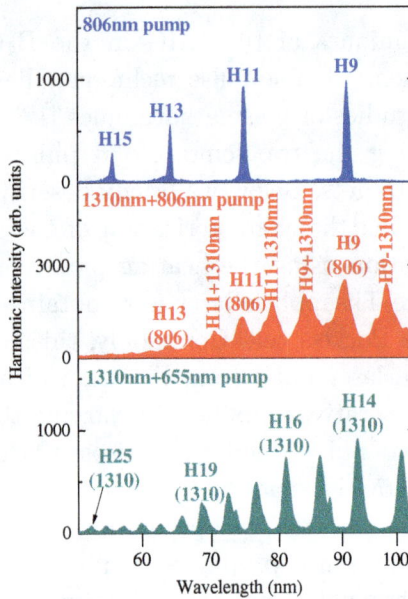

Fig. 5.3. (a) Harmonic spectrum during propagation of double pulses in the vicinity of B_4C target, without the preliminary formation of plasma plume using the 370 ps pulses (thick curve). Harmonic spectrum from B_4C ablation using the 806 nm + 403 nm pump in the case of preliminary formation of plasma plume by the 370 ps pulses (thin curve). The thick curve was multiplied by a factor of 10 for better comparison with the thin curve. These harmonic spectra were obtained at the fluence of $F = 1.2\,\mathrm{J\,cm^{-2}}$. (b) Harmonic spectra obtained using the 806 nm (upper panel), 1310 nm + 806 nm (middle panel), and 1310 nm + 655 nm (bottom panel) pulses propagating through the preliminary formed plasma plume using the 370 ps pulses. Reproduced from [8] with permission from Elsevier.

and the optimization of the HHG using the spatial adjustment of two beams and target have shown the advantages of this technique, which allow the simplified analysis of the high-order nonlinear optical properties of the media, and could be useful for the laser ablation induced HHG spectroscopy.

Another technique for improvement of HHG is the application of the two-color pump technique for the enhancement of harmonic yield and the extension of harmonic cut-off. The commonly used method here is the application of the 800-nm-class lasers and their second-order harmonics (H2) leading to doubling of the number of generated harmonics from odd orders, in the case of single-color pump, to odd and even orders, in the case of TCP. TCP of gases and plasmas allowed the analysis of various microscopic and macroscopic effects, polarization-related processes, and regimes of interaction of the commensurate and incommensurate waves [13–19]. In the following study the TCP scheme comprising 806 nm pulses and second harmonic radiation (403 nm) was used. The 0.5-mm-thick BBO crystal (type I, $\theta = 21°$) was installed inside the vacuum chamber on the path of the focused driving pulse (Fig. 5.1). The conversion efficiency of H2 pulses was 5%. Figure 5.3(a) (thin curve) shows the TCP-induced harmonic spectrum from B_4C plasma. Though the ratio of assistant field (403 nm) compared with fundamental field was notably small (1:20), the use of TCP allowed the three-fold growth of the conversion efficiency of odd harmonics compared with the case of SCP (not shown in the figure). Approximately equal odd and even harmonics were achieved. The thick curve in this figure showing harmonic spectrum using double-pulse scheme was magnified by a factor of 10 for better visibility of the prevalence of the TCP over above-mentioned scheme.

5.1.5. *Different schemes of the two-color pump of plasma*

The above studies were carried out using the commensurate waves, i.e. waves representing some integers from each other. Recently, application of incommensurate waves for the TCP of gases was introduced and showed the advantages in harmonic yield and in generation of individual attosecond pulses using the multi-cycle sources [19]. In the discussed studies [8], SCP (806 nm) and the TCP were compared when the 806 nm radiation was mixed with the radiation (1310 nm) from the optical parametric amplifier. The 1310 nm pump from optical parametric amplifier and its second harmonic for similar experiments was also used. In the case of incommensurate waves, the ratio of the energies of 1310 nm and 806 nm pumps was 1:5, while in the case of commensurate waves, the ratio of assistant field (655 nm) and fundamental field (1310 nm) was 1:4, both of them much larger than the one in the case of the 403 nm and 806 nm pumps (1:20) used in above described experiments.

Figure 5.3(b) presents the SCP spectrum from B_4C plasma (upper panel), as well as two spectra obtained using the TCP comprising the 1310 nm + 806 nm (middle panel) and 1310 nm + 655 nm (bottom panel) configurations of interacting waves. Middle panel shows that the spectrum of generated radiation comprised the sum and difference frequencies of the 1310 nm pulses and harmonics of 806 nm pulses. In particular, one can see the sum ($E_{H9} + E_{1310nm} = 14.79$ eV, $\lambda = 83.9$ nm) and difference ($E_{H9} - E_{1310nm} = 12.89$ eV, $\lambda = 96.1$ nm) waves, alongside the H9 of stronger 806 nm wave ($\lambda = 89.6$ nm). The efficiency of harmonic emission in that case was larger compared with the case of SCP (806 nm), as one can see by comparing the Y-axes of the upper and middle panels, while the cut-off energy in the former case was decreased. Bottom panel of this figure shows the conventional TCP using the commensurate waves (1310 nm + 655 nm). Though the cut-off energy in that case was extended up to the H25 of 1310 nm radiation, the harmonic yield was smaller compared with the case of incommensurate TCP.

The harmonic emission in the case of using the focusing mirror in the XUV spectrometer was collimated along the vertical axis. In that case the raw images of harmonic spectra represented the series of "dots" corresponding to the distribution of harmonics along the XUV region. This method of harmonic images collection allows the observation of generated XUV radiation to be improved. However, it does not allow the analysis of the influence of different parts of driving beam on the spatial distribution of harmonics. To visualize the spatial shapes of harmonics along the vertical axis one has to apply the plane mirror instead of the focusing one. Below we show the images of such harmonic spectra and their analysis in the case of over-excited and optimal plasmas and at the conditions of phase-mismatch along the axis of driving beam (see also Chapter 4). The aim of these studies was to analyze the variable spatial components of harmonics appearing during frequency conversion in the B_4C plasma. Notice that the raw images of spectra presented below are saturated for better visibility, though the line-outs of harmonic spectra were taken from the unsaturated images.

The inset in Fig. 5.4(a) shows the raw image of the H12 to H26 using the 1310 nm + 655 nm pump of the over-excited B_4C plasma using the above-described method of spectrum collection allowing the observation of the spatial shape of harmonic emission in the on-axis and off-axis regions of driving beam propagation. One can see that the on-axis parts of harmonics were significantly suppressed compared with the off-axis components of the same orders. Actually, most of the energy of harmonics was concentrated in the off-axis parts of harmonic beams. The ratio of the on-axis component with regard to the off-axis one ($\sim$1:4) was approximately the same along the whole range of harmonic generation. The spatial distribution of single harmonic (H14) is shown in the bottom panel of Fig. 5.4(a). The reason for this unusual spatial shape of high-order harmonics is related with the formation of the phase mismatching conditions in the axial region of propagation of the driving beam. Large electron density and higher axial intensity of driving beams caused the phase mismatch and self-defocusing. Both these processes lead to deterioration of the optimal conditions of harmonic generation, especially in the central part of the focused beam. Strong intensity of laser radiation (2×10^{14} W cm^{-2}) led to the appearance of a large amount of tunneled electrons during ionization by the driving pulses. Smaller intensity of the driving beam on the wings of the spatial distribution caused lesser influence of above-mentioned impeding processes on the HHG, which led to appearance of the higher-order harmonics dominantly on the off-axis area.

1310nm+655nm pump

H26   H15   H12

(a)

(b)

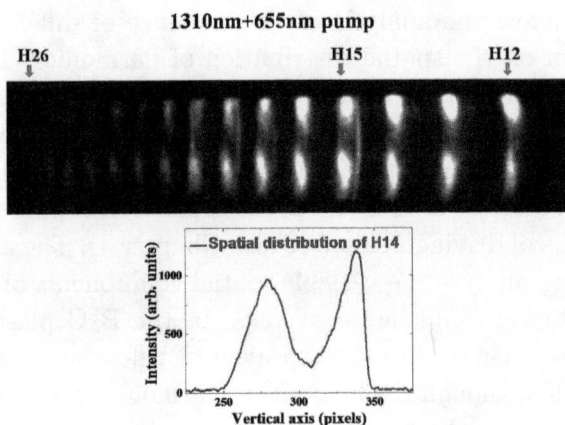

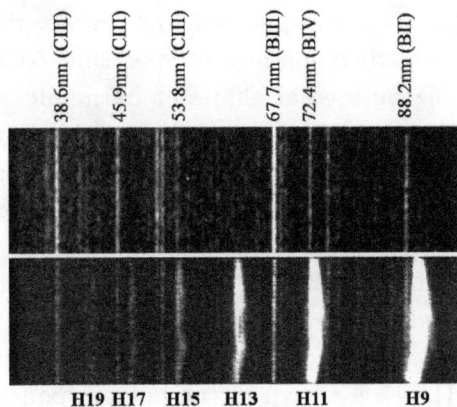

Fig. 5.4. (a) Spatial shapes of H12–H26. Raw image shows the harmonic spectrum in the case of TCP (1310 nm + 655 nm) at the 2×10^{14} W cm^{-2} intensity of driving radiation in the plasma area using the fluence of heating pulse of $F = 1.3$ J cm^{-2}. Bottom panel shows the intensity distribution of H14. (b) Raw images of harmonic spectra using the single-color pump (806 nm) of boron carbide plasma in the case of 0.2×10^{14} (upper pattern) and 2×10^{14} W cm^{-2} (bottom pattern) intensities of driving radiation. Reproduced from [8] with permission from Elsevier.

5.1.6. *Influence of ionic resonances on the harmonic yield*

Symmetry prevents the emission of even harmonics in center-symmetric systems if a single-color driving beam is employed. This restriction was lifted when bichromatic driving fields are employed. The addition of harmonic components has an additional advantage since they permit the exploitation of a broader range of resonances in the plasma species. In the discussed study (i.e. during HHG in the ablated B_4C), the enhancement of single harmonic due to resonance-induced growth of conversion efficiency was not observed, though this process has frequently been observed in other plasma plumes. The coincidence of the wavelengths of harmonics and emission lines of boron carbide did not lead to the growth or decrease of the intensity of these harmonics. The bottom panel of Fig. 5.4(b) presents the raw image of the harmonic spectrum from B_4C plasma in the case of single-color pump (806 nm). One can see the coincidence of the H9, H11, and H15 with the ionic lines of B II (88.2 nm), B IV (72.4 nm), and C III (53.8 nm) respectively, together with the absence of the enhancement of these harmonics. The whole spectrum showed a gradual decrease of harmonic intensity with the growth of harmonic order. The process of resonance-induced enhancement requires the influence of the ionic transitions possessing large oscillator strength on the high-order nonlinear optical response of plasma medium. Two spectra of Fig. 5.4(b) were obtained at the SCP of over-excited B_4C plasma using the relatively weak ($0.2 \times 10^{14}\,W\,cm^{-2}$, upper panel) and strong ($2 \times 10^{14}\,W\,cm^{-2}$, bottom panel) laser intensities. One can see that a 10-fold decrease of the intensity of driving radiation led to the entire disappearance of the harmonic lines from the emission spectrum. Note a homogeneous spatial distribution of harmonics along the vertical axis observed in that case, which pointed out the optimal phase matching conditions for all harmonics along the vertical axis [i.e. for both on- and off-axis components of harmonics, contrary to the case shown in Fig. 5.4(a)].

5.1.7. *Discussion of experiments*

Below we address some findings of these studies.

(1) In this study, we analyzed the comparison of the HHG using different TCP schemes (i.e. using the commensurate and incommensurate groups of waves). Actually, three TCP schemes were used. The results of 806 nm + 403 nm induced HHG are presented in Fig. 5.3(a), while results of using the MIR + 806 nm and MIR + H2 pumps are shown in Figs. 5.3(b) and 5.4(a). The relative comparison of these three schemes showed the best conversion efficiency in the case of former pump, probably due to shorter wavelengths of the sources of this TCP scheme. Meanwhile, second scheme allows the denser spectrum of harmonics consisting on the sum and difference components of interacting waves to be produced. The third scheme showed intermediate results among these TCP schemes from the point of view of the HHG conversion efficiency.

(2) The spatial resolved TCP studies [Fig. 5.4(a)] show the dominant off-axial distribution at "non-optimal" conditions of plasma formation and high intensities of pump waves, which was attributed to the influence of electrons on the phase matching conditions. One can also speculate that this redistribution of the harmonic yield could also be explained by the changed phase matching conditions of the long/short trajectories of accelerated electrons. However, we do not think that our observation was related with the latter process. In the additional experiments using different ablated targets [see also the bottom panel of Fig. 5.4(b)], the homogeneous off- and on-axis components of harmonics were observed, which demonstrate the equality of conditions for the long and short trajectories of accelerated electrons leading to harmonic generation. Thus the difference of the phase matched conditions in the cases of long and short trajectories was insignificant in the case of "optimally prepared plasma" and "optimal intensity" of converting pulses. The prevalence of the off-axis harmonics over on-axis ones appeared only in the case of high plasma densities and high on-axis intensities of the driving pulses.

(3) As we wrote in the previous subsection, the lower conversion efficiency in B_4C plasma, compared with graphite plasma, could be explained by possible participation of B_4C molecules in HHG, contrary to atoms and clusters of carbon. There are also other assumptions to explain the lower conversion efficiency in B_4C plasma, e.g. the lower density of B and C atom/ions in the B_4C plasma plume than in the boron and graphite plasmas, and/or the different phase matching factors in these constituencies. In particular, as one can see in Fig. 5.2, the plasma emission may indicate the ratio of B/C atom/ions in different plasma plumes, and the further comparison of harmonic efficiency could give a hint on whether neutral molecules decisively contribute to HHG or not.

As it has already been mentioned, the most probable reason for the difference of harmonic emission properties of B_4C and two other ablated targets (boron and graphite) is the different constituencies of plasmas in these two pairs (B_4C-B and B_4C-C) under the optimal conditions of plasma formation. Regarding the comparison of B_4C and B plasmas, the stronger conversion efficiency in the former plasma was observed, which contradicts the above-mentioned assumption. This experimental observation points out the larger density of emitters in the case of boron carbide plasma compared with boron plasma. As for comparative results of B_4C and C plasmas, one can assume that the specific properties of latter plasma formation allow formation of carbon nanoparticles. The response of carbon plasma could thus be a sum of atomic/ionic and nanoparticle harmonic emitters, which prevail over harmonic emission from the B_4C plasma. The ablation of the latter target does not lead to formation of clusters, which was confirmed during TEM measurements of the debris of B_4C ablation. The second option is the difference between the phase matching conditions for these three groups plasmas. This assumption could be true once one analyzes the influence of each of constituencies on the waves propagation. One cannot rule out this opportunity. However, the lack of knowledge of the spectral dispersion of these components does not allow one to even make qualitative estimates of the difference of the wave vectors of interacting radiation and harmonics in these three components (B_4C, B, and C).

The quantitative definition of the relative ratio of B/C atom/ions in different plasma plumes using the plasma emission spectra shown in Fig. 5.2 at similar fluence of ablating pulses could be true once we use the same fluence of heating pulses for the harmonic generation experiments. Actually, the fluencies of heating pulses for the HHG experiments were other than for the plasma emission studies. In each case the "optimal" fluence of heating pulses was chosen. This term refers to the maximal efficiency of harmonic generation in the studied plasmas plumes. What one can derive from Fig. 5.2 is the difference in the incoherent emission under similar conditions of ablation. We again reiterate that these conditions are unsuitable for efficient HHG. To conclusively define whether neutral molecules contribute or not for HHG one has to use the time-of-flight mass spectrometry methods.

(4) It was found during comparison of the HHG from the B_4C, B, and C that the boron carbide plasma allows less conversion efficiency than graphite. Meanwhile the HHG in that case is more stable, so the B_4C plasma can be considered as a best choice among three ablated targets from this point of view. The stability of the harmonics from three plasmas was compared by analyzing the relative deviation of the harmonic yield from the averaged values. It was found that, with the growth of the shots of ablating beam on the same spot of targets, the harmonic yield started to decrease due to the modification of the target surface. This modification was less influential in the case of boron carbide, which was confirmed by analyzing the target surface after a few thousands shots on the used samples. To improve the stability of harmonics one can drag up and down the target during HHG experiments to keep the stable yield of harmonics.

Notice that there are various characteristics which could be prevailed once one chooses the target for ablation and HHG. In this particular case three characteristics were compared: (a) stability of harmonic emission, (b) extension of harmonic cutoff, and (c) harmonic yield. The B_4C plasma was considered as the best choice among three ablated targets from the point of view of factor (a), while the B plasma showed best characteristics from the point of view of factor (b), and the graphite plasma showed best characteristics from the point of view factor (c).

5.2. Two-color High-order Harmonic Generation in Plasmas: Efficiency Dependence on the Generating Particle Properties

The main drawback of HHG is very low conversion efficiency. As already stated, the use of a driving field consisting of the laser field and its second harmonic is one of the reliable methods for enhancing the harmonic yield. The HHG using a two-color pump was experimentally studied in gas media using parallel [20–23] or orthogonal [20, 24–27] polarization configurations. Numerous theoretical studies [28–37] addressed different aspects of two-color HHG. In particular, they showed that the role of the second harmonic field is essentially different in the cases when its intensity is much lower than that of the laser field and when the two intensities are comparable. Though more complicated experimentally, the latter case allowed for HHG with very high conversion efficiency (5×10^{-5} for the 38th harmonic [24]). The first explanation of strong HHG enhancement in two-color orthogonally polarized fields with comparable intensities of the components was given in [20, 24] in terms of the formation of a transient linear polarization of the field, selection of a short quantum path component, which has a denser electron wave packet, and higher ionization rate compared with the single-color pump case.

The radiation of 800-nm-class lasers combined with their second harmonic was used at the initial stages of those studies. Various schemes, among which are the generation of the 400 nm wave in a separate channel with further mixing with the 800 nm wave in gases, as well as direct generation of second harmonic (H2) in thin BBO crystals followed with focusing of two beams in the converting medium, were introduced [23, 25]. Experimental [26, 27] and theoretical studies [34] also demonstrated strong dependence of the HHG yield on the relative phase between the fundamental and second harmonic radiation.

The macroscopic aspects of two-color HHG were addressed already in early theoretical studies [38, 39] and have received much attention very recently [40, 41]. In general, the phase-matching problem in the case of a two-color field is more complicated than in the single-color case because it requires taking into account not only the phase difference between the pump field and the generated one but also the phase difference between the two pump fields which changes during propagation.

The TCP approach has been applied to harmonic generation in laser-produced plasmas as well. The phase matching is especially important for the plasma HHG because the free electrons' density is comparable with that of the emitters, potentially leading to non-negligible phase mismatch even for relatively low-density plasmas. The first observation of an increase in the harmonic generation efficiency from narrow ($\sim$0.3-mm-long) plasma plumes irradiated by an intense two-color orthogonally polarized fields was reported in [42]. The intensity of the second harmonic was weak (the energy ratio of the laser pulse and H2 was 50:1).

The next stage of two-color HHG is based on the application of mid-infrared pulses from optical parametric amplifiers combined with their second harmonics. The MIR+H2 scheme, particularly 1300 nm + 650 nm, was successfully applied to both gas [27, 34] and plasma [43] HHG. The higher conversion efficiency into the second harmonic in the mid-infrared region allows for the experimental study of two-color HHG for comparable intensities of the driving field components.

In this section, we analyze two-color HHG in an extended (5-mm-long) laser-produced plasma using 1310 nm emission of optical parametric amplifier and its second harmonic of comparable intensities [44]. A significant enhancement of high-order harmonics compared with the SCP was observed; in particular, the highest-order harmonics are very pronounced in the two-color case whereas they are hardly detectable in the single-color case. The increase of the HHG efficiency in TCP was compared experimentally for different plasmas (silver, gold, and zinc) and different harmonic orders. The experimental results are analyzed theoretically. In particular, it was found that the commonly used HHG theory, which assumes the 1s ground state of the generating particle, cannot reproduce the experimentally observed HHG spectrum. It was therefore suggested that a theory using actual quantum numbers for the ground states of the species used for HHG (silver, gold, and zinc ions in our case) should be developed. Having in mind a significant role of the propagation effects in the extended plasma target, an approach to calculate the macroscopic HHG signal was suggested taking into account the variation of the phase difference between the frequency components of a driving field along the propagation distance. The integrated macroscopic signal calculated using the improved micro-scopic response reproduces the measured essential dependence of the enhancement on the plasma type, as well as on the harmonic order. Moreover, the theoretical studies allow for an important aspect of the HHG enhancement in a two-color field which was not discussed earlier to be clarified. Namely, according to the simple-man HHG model, in the single-color field the majority of the ionized electrons come back to the parent ion with zero energy; it was shown in the discussed work that in the two-color orthogonally polarized field with comparable intensities of the components, the majority of the electrons can come back with energy corresponding to the cut-off harmonics. The conditions for which this takes place were found.

5.2.1. *HHG using single-color and two-color pumps*

The scheme similar to the one described in section 5.1 was used in these studies (Fig. 5.5). The signal radiation was used as the driving pulse for HHG in LPP. The intensity of the 1310 nm pulses focused by a 400 mm focal length lens inside the extended plasma was $\sim 2 \times 10^{14}$ W/cm^2. The driving beam was focused, with the focal spot size of 90 μm, into the prepared extended plasma at a distance of ~ 100 μm above the target surface.

The HHG efficiency in the used LPP (Ag, Au, and Zn) was optimized with respect to the delay between the heating and driving

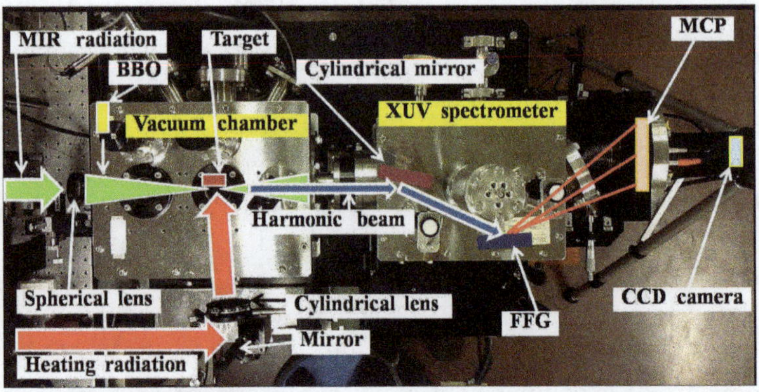

Fig. 5.5. Experimental setup for harmonic generation in metallic plasma. Reproduced from [44] with permission from Optical Society of America.

pulses. The harmonic generation efficiency abruptly increased once the delay exceeded 5 ns. In the case of the used particles (with atomic weights of 65, 108, and 197 for Zn, Ag, and Au, respectively), the maximal harmonic yields were observed at 25, 35, and 45 ns delays due to different velocities of the ablated particles. At longer delays for each specific sample, for a fixed distance between the target and the laser beam ($\sim$150 μm), the harmonic yield gradually decreased until it disappeared entirely at $\sim$150–200 ns.

The density characteristics of the LPP at the delays of 25–45 ns were estimated using the HYADES code [45]. For the heating pulse intensity of 2×10^9 W/cm^2, the ionization level and the ion density of the silver plasma were estimated to be 0.4 and 2×10^{17} cm^{-3}, respectively.

The harmonic yield from Ag, Au, and Zn plasmas using MIR pulses (1 mJ, 1300 nm) was significantly weaker compared with the 7-mJ, 806-nm pump. Such a decrease of the high-harmonic signal with the driving wavelength was shown earlier in gases [46, 47]. The harmonic cutoff energy in the case of MIR pulses was also lower compared with that in the 806-nm case, contrary to the expectations of the cutoff extension, due to very low conversion efficiency, which did not allow for the observation of harmonics with wavelengths below 50 nm. However, the situation became completely different in the TCP case.

The 0.5-mm thick barium boron oxide (BBO) crystal ($\theta = 21°$) was installed inside the vacuum chamber in the path of a focused signal beam (see Fig. 5.5) to generate its second harmonic. The efficiency of conversion into the ~650-nm radiation was ~27%. Two pulses overlapped both temporally and spatially in the extended plasma and allowed for a significant enhancement of odd harmonics, as well as generation of even harmonics with intensity similar to that of the odd ones.

The focusing conditions of TCP also influenced the HHG efficiency in LPP. The optimal focusing condition of HHG was found by varying the focal length of the focusing spherical lens. Maximum harmonic yield was observed when the confocal parameter of MIR beams was approximately equal to the length of the plasma plume (~5 mm).

The absolute value of the energy of harmonics generated from the silver plasma was measured using the method described in [48]. Briefly, the method is based on the measurements of calibrated signals in the ultraviolet and XUV ranges. The measured energy of the harmonics generated in the 40-nm spectral region is $0.04 \, \mu J$ in the case of Ag plasma. Taking into account the 1-mJ energy of 1310 nm pulses, this gives 4×10^{-5} conversion efficiency for these harmonics.

A significant enhancement of the harmonic yield in the case of TCP was observed in all the metal plasmas used. Silver plasma allowed for generation of harmonics up to H23 in the case of 1310 nm pump, see Fig. 5.6(a), thick red curve. The TCP (1310 nm + 655 nm) allowed for odd and even harmonic generation of orders up to the 40 s (thin blue curve). The same tendency was observed in the gold plasma, see Fig. 5.6(b): the harmonics generated using TCP were stronger than those in the SCP case, the harmonic cutoffs were extended up to the orders of the 40 s and 50 s. The enhancement factor depended on the spectral range and varied from a few units to several tens. Zn plasma demonstrated similar features, see Fig. 5.6(c).

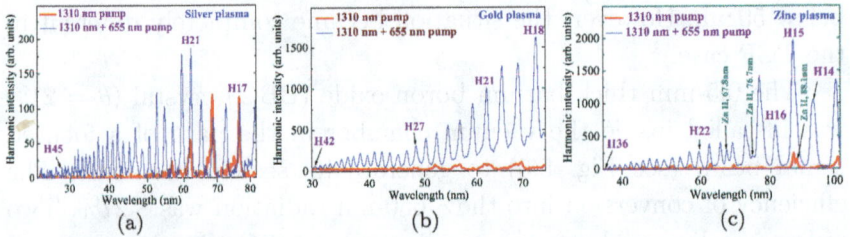

Fig. 5.6. HHG in (a) Ag, (b) Au, and (c) Zn plasmas using 1310 nm and 1310 nm + 655 nm pumps. Reproduced from [44] with permission from Optical Society of America.

The enhancement factor in the case of TCP compared with SCP significantly varied in different ranges of harmonic spectra and also depended on the plasma species. In the case of Ag plasma, the enhancement factor for H21–H23 was in the range of 3 to 5. For higher orders, the pronounced harmonic peaks did not appear in the case of SCP in the experimental conditions used in these studies, whereas introduction of BBO in the path of the driving beam led to a drastic change of emission pattern in this spectral range, corresponding to at least approximately 50-times growth of harmonic emission of orders of the thirties to forties. Similar behavior of the enhancement factor was observed in the case of gold plasma, though the value of this parameter was smaller compared to the case of silver plasma both in the longer- and shorter-wavelength regions. Finally, zinc plasma allowed observation of 5- to 10-fold growth of the conversion efficiency into the lower-order harmonics in the case of TCP, whereas for higher-order harmonics this factor decreased.

5.2.2. *Microscopic response*

Theoretical approach used in these studies was based on the modification [49, 50] of the original analytical HHG theory [51]. In crossed field geometry, when neglecting the electron magnetic drift in a laser field, the dipole moment of the particle can be written as follows:

$$\mathbf{r}(t) = i \int_0^\infty d\tau \left(\frac{\pi}{\varepsilon + i\tau/2} \right)^{\frac{3}{2}} \left\langle \mathbf{d} \left(\mathbf{p}_{st} - \frac{\mathbf{A}(t-\tau)}{c} \right), \mathbf{E}(t-\tau) \right\rangle \mathbf{d}^*$$

$$\times \left(\mathbf{p}_{st} - \frac{\mathbf{A}(t)}{c} \right) \times \exp \left[-iS(\mathbf{p}_{st}, t, \tau) - \int_0^t \frac{W(t')}{2} dt' \right.$$

$$\left. - \int_0^{t-\tau} \frac{W(t')}{2} dt' \right] + c.c., \tag{5.1}$$

where c.c. denotes complex conjugation; $\mathbf{E}(t)$ and $\mathbf{A}(t)$ are the total electric field and the total vector potential of a two-color laser pulse, respectively; c is the speed of light; ε is a small regularization parameter; τ is the time of the electron's free motion in the continuum;

$$\mathbf{p}_{st}(t, \tau) = \frac{1}{\tau} \int_{t-\tau}^t \frac{\mathbf{A}(t')}{c} dt' \tag{5.2}$$

is the stationary point of the quasiclassical action $S(\mathbf{p}, t, \tau)$; $\mathbf{d}(\mathbf{p})$ is the dipole matrix element corresponding to the transitions from the bound state to the continuum plane wave; $W(t)$ is the time-dependent ionization rate. Expressions for $S(\mathbf{p}, t, \tau)$ and $W(t)$ used for calculations can be found in [50].

In all calculations of this research the two components of a laser field are orthogonal and have the following shape:

$$\begin{cases} E_x(t) = E_1(t) = E_0 \exp \left[-\ln(4) \frac{t^2}{T^2} \right] \exp(-i\omega t) + c.c. \\ \\ E_y(t) = E_2(t) = \sqrt{\alpha} E_0 \exp \left[-\ln(4) \frac{t^2}{T^2} \right] \exp(-i2\omega t + i\varphi) + c.c. \end{cases}$$

$$\tag{5.3}$$

where E_0 is the amplitude of the main component of the laser field (in all the calculations the fundamental intensity is 10^{14} W cm^{-2} unless stated otherwise explicitly); α is the ratio of the intensities of the laser field components; $\omega = 0.035$ atomic units corresponding to the fundamental wavelength of 1300 nm; T is the full width at half-maximum intensity, which is taken equal to 16 cycles; φ is the relative phase. Calculations were performed for various α and φ and for several ionic targets (Ag^+, Au^+, and Zn^+). Note that not only the ionization potential I_p, but also the dipole matrix element $\mathbf{d(p)}$ is specific for every target. A general expression for the dipole matrix element corresponding to the transition was derived from an arbitrary nonmagnetic bound state of the hydrogen-like atom to the continuum plane wave:

$$
\mathbf{d(p)} = \frac{i^{l-1} 2^{2l+3/2} \, (l+1)! \, \gamma^{l+3/2}}{\pi \sqrt{n}} \sqrt{(n-l-1)! \, (n+l)! \, (2l+1)}
$$

$$
\times \nabla_{\mathbf{p}} \left[p^l P_l \left(\frac{p_z}{p} \right) \sum_{m=0}^{n-l-1} \frac{(2\gamma)^m}{m! \, (2l+1+m)!(n-l-1-m)!} \right.
$$

$$
\left. \times \frac{d^m}{d\gamma^m} \frac{\gamma}{(\gamma^2 + p^2)^{l+2}} \right], \tag{5.4}
$$

where n is the principal quantum number; l is the azimuthal quantum number; z indicates the quantization axis; P_l is the Legendre polynomial; $\gamma = q/n$, q is the core effective charge (in all the calculations $q = 2$).

Figure 5.7(a) shows spectra (nm-scale is used here to simplify the comparison with Fig. 5.6) of silver ion response obtained with the same $I_p = 21.48$ eV and various expression for $\mathbf{d(p)}$. It is clearly seen that the particular form of the dipole matrix element strongly influences the shape of the harmonic spectrum. Note also that the HHG spectrum obtained with the widely used expression for $\mathbf{d(p)}$ with $n = 1$ and $l = 0$ differs essentially from the experimental one, see Fig. 5.6(a), thin blue line. In contrast to that, the shape of the spectrum obtained with the actual quantum numbers of the outer electron in Ag^+ ($n = 4$ and $l = 2$) is quite close to the shape of the experimental one. Thus, the proper choice of the expression for $\mathbf{d(p)}$ is crucial for the correct reproduction of the HHG spectrum in theoretical calculations. The provided analysis is based on the microscopic response, however, in the macroscopic section of the paper it will be shown that the shape of the spectrum after propagation through the medium in a two-color case is quite close to the shape of the microscopic one obtained with $\varphi = \pi/2$. Therefore, the conclusions on the effect of the matrix element on the shape of the generated harmonic spectrum are valid for the macroscopic response of the medium also. For better agreement with the experimental results, in all further calculations the actual quantum numbers of the outer electron and the actual ionization potentials of the target ions are used (which are $n = 4, l = 2, I_p = 21.48$ eV for Ag^+; $n = 5, l = 2, I_p = 20.5$ eV for Au^+; $n = 4, l = 0, I_p = 17.96$ eV for Zn^+) unless stated otherwise explicitly.

The calculations were performed with the step $\pi/24$ for the variable φ, for several α and for intensities of the fundamental field in the range of $5 \times 10^{13} - 2 \times 10^{14}$ W cm^{-2} for all the targets. The results for various α and various intensities of the fundamental field are very similar qualitatively and show that for all the targets the addition of the second field leads to an increase of the harmonic yield. The best agreement with experimental results in terms of the width of the harmonic spectra is observed for the intensity of the fundamental field equal to 10^{14} W cm^{-2}. Figures 5.7(b) and 5.7(c) show the spectra $|r_\Omega(\varphi)|^2 = \left| \int_{-\infty}^{\infty} \ddot{r}(t) \exp(i\Omega t) \, dt \right|^2$ of the silver ion response for this intensity and several α and φ. Figure 5.7(b) shows the worst case $(\varphi = 0)$ and Fig. 5.7(c) shows the best one

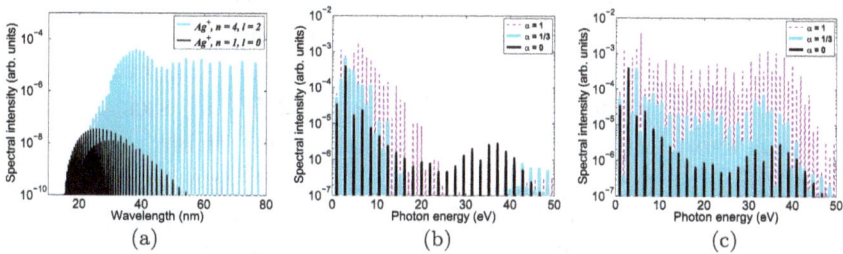

Fig. 5.7. Spectra of a silver ion response: (a) with $\alpha = 1/3, \varphi = \pi/2$ and various $\mathbf{d(p)}$ (see legend); (b) with $\varphi = 0$ and various α (see legend); (c) with $\varphi = \pi/2$ and various α (see legend). Results in (b) and (c) are obtained with the actual quantum numbers. Reproduced from [44] with permission from Optical Society of America.

($\varphi = \pi/2$). The harmonic spectra obtained with other values of φ are in between these two cases. It is clear that the effect of the second field depends strongly on α and φ, and under optimal conditions the gain in harmonic intensity can be almost 3 orders of magnitude. Results for gold and zinc ions are qualitatively similar, but the gain is lower.

The explanation of this enhancement was given in [13, 24] in terms of the formation of a transient linear polarization of the field, selection of a short quantum path component, which has a denser electron wave packet, and higher ionization rate compared with the SCP case. However, the analysis done in [13,24] is not comprehensive enough to answer all the questions related to the HHG enhancement in TCP. Here we provide a more detailed analysis of the classical electronic trajectories in the two-color laser field.

Due to the orthogonal polarizations of the field components, the electron's motion caused by each of the field components can be treated separately. The electron's motion in the x-direction (due to the fundamental field) in a two-color field is exactly the same as in a single-color one. The motion in the y-direction (due to the second-harmonic field) is synchronized with that in the x-direction at the instant of the electron's release from the particle: both motions start at the origin at the same time with zero velocities. Electron can recombine with the parent ion if it returns close enough to the origin simultaneously in both directions.

Figure 5.8(a) shows the y-coordinate of the electron at the instant of its return to the origin in the x-direction and the width of the returning wavepacket versus the time of ionization. The calculations were made using the formula proposed in [52, 53]. It is clearly seen that for short enough trajectories with the time of birth exceeding ∼0.07 of a laser cycle [as follows from Eq. (5.3), zero time coincides with the maximum of the oscillation of the fundamental field, see also Fig. 5.4(b)] the displacement of the electron in y-direction at the instant of recollision is smaller than the width of the returning wavepacket. It means that only these trajectories contribute to HHG in the case of a two-color field and the longer trajectories' contributions are suppressed. A similar conclusion was made in

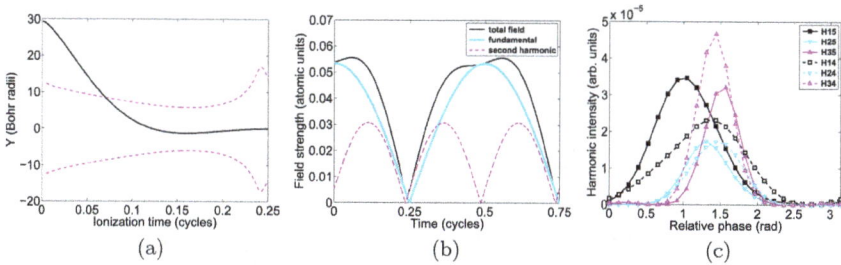

Fig. 5.8. (a) Y-coordinate of the electron at the instant of its return to the parent ion in x-direction (solid line) and the y-width of the returning electron wavepacket (dashed line) as a function of ionization time. The calculations were made using formula proposed in [52,53]; $\alpha = 1/3$, $\varphi = 1.4$. (b) Time dependence of the absolute value of the total field and field components (see legend); $\alpha = 1/3$, $\varphi = 1.4$. (c) Harmonic intensity vs. relative phase φ for several harmonics (see legend); $\alpha = 1/3$. Reproduced from [44] with permission from Optical Society of America.

[13, 24]. However, we would like to note that Fig. 5.8(a) provides more information. The classical picture predicts that the cut-off harmonic is due to the trajectory launched at 0.05 of a laser cycle. It means that almost all short trajectories (0.07 is close to 0.05) can contribute efficiently to HHG under the optimal conditions in the TCP scheme (in contrast to [13, 24], where it was stated that only very short trajectories with travelling time shorter than 0.3 optical cycle contribute to HHG efficiently). That is why the HHG spectrum in the two-color case can be very wide.

Figure 5.8(b) shows the absolute value of the total field in the single-color and the two-color cases for the same conditions as shown in Fig. 5.8(a). One can see that in the two-color case the absolute value of the field is larger. It means that the ionization rate in a two-color field is higher compared with the single-color case due to a very strong dependence of ionization rate on the value of the laser field in the tunneling ionization regime. This increase of the ionization rate was highlighted in [13, 24].

However, neither the quasi-linear polarization of the field between the instants of the electron's release and return (discussed in [13,24]), nor the increase of the ionization rate can completely explain the dramatic increase of the HHG efficiency in a two-color field. Indeed, using linearly polarized single-color field with higher intensity one might benefit from both of these factors. However, this is not the case: the HHG efficiency saturates with laser intensity in such a field. To solve this issue, another important feature of the HHG process in a two-color field should be taken into account. This feature is the redistribution of the electronic density in the continuum electron wave packet: the maximum increase of the absolute value of the field and, consequently, the maximum enhancement of the ionization rate is observed for the time interval corresponding to the birth of the short trajectories responsible for efficient HHG in the two-color case, see Fig. 5.8(a). Summarizing the results shown in Figs. 5.8(a) and 5.8(b) one can explain the strong enhancement of HHG yield in a two-color field as follows. There is a family of electronic trajectories that are not significantly deflected by the second harmonic in the orthogonal direction. Fortunately, these are exactly the trajectories whose weights strongly increase due to the change of the tunnel ionization dynamics in a two-color laser field. As a result, the HHG yield in the TCP case is enhanced significantly. Note also that most of the electrons are released at the maximum of the field. In the single-color case, half of them never come back to the origin and the other half produce only low harmonics in the HHG spectrum (in the SCP case the longest trajectories begin at the field maximum and return to the parent ion with almost zero velocities). As a result, only a small fraction of electrons are responsible for HHG. In the two-color case, the total field maximum is shifted so that all the electrons released near the field maximum return to the origin with high energies: the cut-off harmonic is due to the trajectory launched at 0.05 of a laser cycle, thus very close to the instant when the field is maximal, see Fig. 5.8(b). Thus, in the two-color case a greater part of the released electrons contribute to HHG rather than just leave the vicinity of the parent ion. The efficiency of HHG can be therefore much higher for the TCP case compared with the SCP one even at the same ionization degree of the gas medium.

From Fig. 5.8(b) it is also obvious that the change of the relative phase φ would shift the time interval with maximum enhancement of the field absolute value, hence, another group of trajectories (shorter or longer ones) would predominantly contribute to HHG. Different trajectories correspond to different harmonics in the HHG spectrum, hence, the optimal relative phase should be different for different harmonics. For the odd harmonics generated due to the electron motion in the x-direction this conclusion agrees very well with the results of the numerical calculations presented in Fig. 5.8(c). The even harmonics are generated due to the electron motion in the y-direction, and the optimal relative phase (slightly smaller than $\pi/2$) is the same for all of them.

5.2.3. *Macroscopic response*

As shown in the previous subsection, the microscopic response is very sensitive to the relative phase φ. This phase changes during the fields' propagation in the medium. Thus, although a very pronounced intensity enhancement of the microscopic response at certain relative phases was found in the previous section, this might not eventually cause a comparable enhancement in the macroscopic signal. Therefore, a complete study of the HHG phase matching in a two-color field is necessary to compare the calculated results with the experimental ones.

Let us consider the phase matching for 1D propagation of the fundamental field and the second harmonic along z-axis. The fields inside a nonlinear medium are:

$$E_1(z,t) = E_0(z,t) \exp\left[-i\omega(t - z/c) + i\varphi_1(z)\right] + c.c., \tag{5.5}$$

$$E_2(z,t) = \sqrt{\alpha} E_0(z,t) \exp\left[-i2\omega(t - z/c) + i\varphi_2(z)\right] + c.c., \tag{5.6}$$

where $E_0(z,t)$ is the slowly-varying field envelope. By introducing

$$t' = t - \delta t, \tag{5.7}$$

where $\delta t = \varphi_1/\omega$, the driving fields given by Eqs. (5.5) and (5.6) can be written as

$$E_1(z,t') = E_0(z,t) \exp\left[-i\omega(t' - z/c)\right] + c.c., \tag{5.8}$$

$$E_2(z,t') = \sqrt{\alpha} E_0(z,t) \exp\left[-i2\omega(t' - z/c) + i\varphi_2'(z)\right] + c.c., \tag{5.9}$$

where

$$\varphi_2' = \varphi_2 - 2\varphi_1. \tag{5.10}$$

The nonlinear polarization $P(z, t')$ induced by the fields (5.8) and (5.9) can be written as

$$P(z, t') = \int_\Omega P_\Omega(\varphi_2') \exp\left[-i\Omega(t' - z/c)\right] d\Omega + c.c., \tag{5.11}$$

where $P_\Omega(\varphi_2') = N r_\Omega(\varphi_2')/\Omega^2$; N is the density of the generating particles. The calculation of the microscopic response spectrum $r_\Omega(\varphi_2')$ for given relative phase φ_2' between the driving fields was described in the previous section. Substituting Eqs. (5.7) and (5.10) in Eq. (5.11) one can find

$$P(z, t) = \int_\Omega P_\Omega \left[\varphi_2(z) - 2\varphi_1(z)\right]$$

$$\times \exp\left[-i\Omega(t - z/c) + i\frac{\Omega}{\omega}\varphi_1(z)\right] d\Omega + c.c. \tag{5.12}$$

The intensity of the macroscopic response is (here we assume that the refraction index for the high harmonic field is unity)

$$I_\Omega = N^2 \left| \int_0^L r_\Omega \left[\varphi_2(z) - 2\varphi_1(z)\right] \exp\left[i\frac{\Omega}{\omega}\varphi_1(z)\right] dz \right|^2. \tag{5.13}$$

Considering the dependence of the phases φ_2 and φ_1 on z we take into account the presence of the initial relative phase δ:

$$\varphi_1(z) = z\omega(n_1 - 1)/c, \tag{5.14}$$

$$\varphi_2(z, \delta) = z2\omega(n_2 - 1)/c + \delta, \tag{5.15}$$

where n_1 and n_2 are the refraction indices for the fundamental field and the second harmonic.

The macroscopic response after the averaging over the initial relative phase (in our experimental conditions this phase randomly changed from shot to shot) is found to be

$$\bar{I}_\Omega = \frac{1}{\pi} \int_0^\pi d\delta N^2 \left| \int_0^L r_\Omega \left[\varphi_2(z,\delta) - 2\varphi_1(z) \right] \exp \left[i\frac{\Omega}{\omega} \varphi_1(z) \right] dz \right|^2,$$

(5.16)

where the phases φ_2 and φ_1 are given by Eqs. (5.14) and (5.15). One can assume that

$$n_1 - 1 = 4(n_2 - 1).$$

(5.17)

This is the case for the plasma dispersion under the conditions when the free electron dispersion is described within the Drude model and the atomic or ionic dispersion is neglected; the latter assumption is valid for the ionization rate exceeding 10–20%, thus, it is definitely applicable to our experimental conditions. Moreover, Eq. (5.17) also describes the geometrical dispersion for a certain focusing.

The theoretical results for the behavior of the macroscopic signal versus the propagation distance in single- and two-color cases are shown in Fig. 5.9. One can see that the signal of the qth harmonic in the single-color case goes down to zero after propagation over

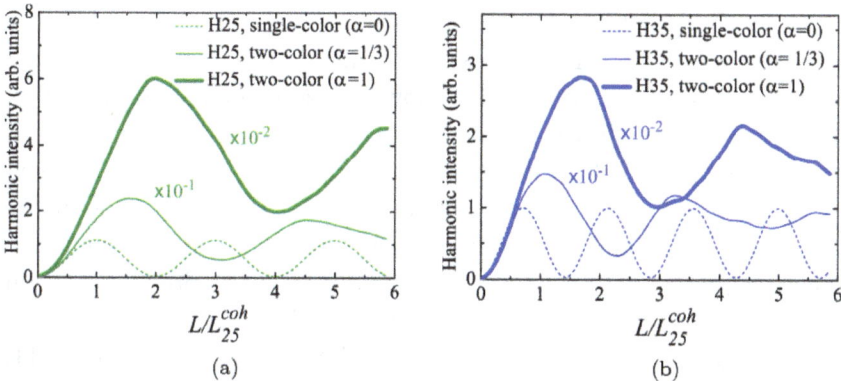

Fig. 5.9. The macroscopic signals of (a) H25 and (b) H35 as functions of the propagation distance in the single-color (dotted line) and two-color (solid lines) cases calculated for silver plasma. The signals for the two-color fields with $I_{2\omega} = I_\omega/3(\alpha = 1/3)$ and $I_{2\omega} = I_\omega(\alpha = 1)$ are multiplied by 10^{-1} and 10^{-2}, respectively. Reproduced from [44] with permission from Optical Society of America.

the doubled coherence length L_q^{coh}. In contrast, the signal in the two-color case does not diminish to zero. This is a consequence of strong dependence of the microscopic response (both its amplitude and phase) on the relative phase in the integral (5.13). Indeed, due to this dependence only a limited range of the relative phases (and, accordingly, a limited range of z) contributes to the integral, preventing vanishing of the integral which would be the case if the variation of the phase in the exponent was 2π.

The calculations of the macroscopic signals as functions of the propagation length for Zn^+ and Au^+ in the two-color case show that their behavior is qualitatively similar for all the conditions considered in our calculations: the signal of a certain harmonic reaches a maximum at approximately 1.5–2 coherence length for this harmonic (calculated for a single-color case) and then oscillates but does not go down to zero for the considered propagation length.

In Fig. 5.10, the spectra calculated for a given propagation length for different plasma species are presented. Figure 5.10(a) shows that the shape of the spectrum after propagation through the medium in a two-color case is quite close to the shape of the microscopic one obtained with $\varphi = \pi/2$, see Fig. 5.7(c). The reason for this similarity is the strong dominance of the response from the regions with the optimal phase delay in the total response of the medium. Another consequence of this fact is that the signal of a certain harmonic in a two-color case does not go down to zero for any propagation length, as it was shown before in Fig. 5.9. From the comparison of Figs. 5.6(a) and 5.10(a) it is obvious that a good agreement between experimental and theoretical results can be achieved only with the use of the actual quantum numbers of the outer electron. Thus, the macroscopic analysis confirms the conclusion in the previous section about the crucial role of the expression for $\mathbf{d}(\mathbf{p})$.

The gain due to the two-color field is presented in Fig. 5.11. One can see that for given α the highest gain is achieved for silver, in agreement with the experiment. For $\alpha = 1/3$ the typical gain obtained in the calculations for silver is 10–30, which is in reasonable agreement with the experimental value (about 50). Comparing the results for the same medium (silver) and different intensities of the

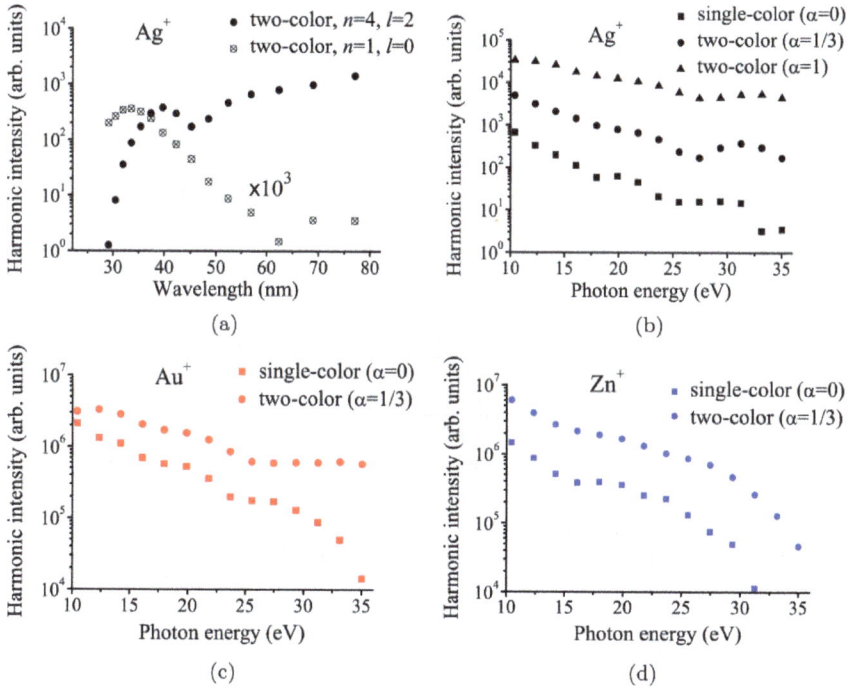

Fig. 5.10. The macroscopic spectrum calculated for (a)–(b) silver, (c) gold, and (d) zinc plasma in the single- and two-color cases for a given propagation distance $L = L_{21}^{coh}$. Results in (a) are obtained with $\alpha = 1/3$ and various $\mathbf{d}(\mathbf{p})$ (see legend). Reproduced from [44] with permission from Optical Society of America.

second harmonic (namely, for $\alpha = 1/3$ and $\alpha = 1$) one can see that the gain grows dramatically with increasing α: for $\alpha = 1$ it is approximately an order of magnitude higher than for $\alpha = 1/3$. This very pronounced dependence of the gain on α can explain some disagreement between the theoretical results for $\alpha = 1/3$ and the experimental ones, having in mind the limited accuracy of experimental measurement of α and possible modulation of this quantity in time and space.

Moreover, from Fig. 5.11 one can see that the enhancement in silver plasma is more pronounced, in agreement with the experiment, see Fig. 5.6(a). For gold and zinc the enhancement for different harmonics is more similar. Experimentally this is the case for

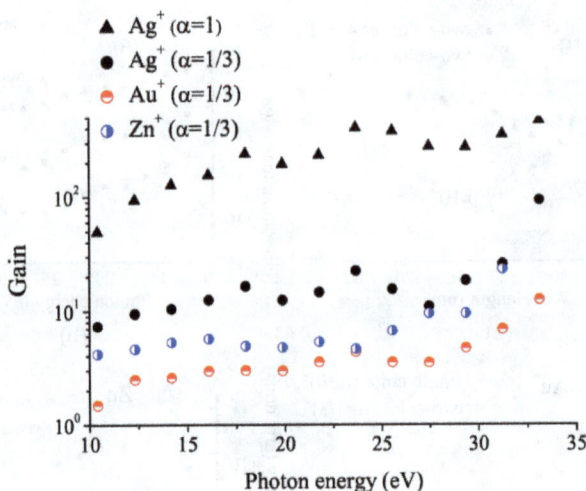

Fig. 5.11. The gain in harmonic intensity in the TCP case compared to the SCP one for different plasmas. Reproduced from [44] with permission from Optical Society of America.

gold, see Fig. 5.6(b), although for zinc even some decrease of the enhancement is observed, see Fig. 5.6(c).

5.3. High-order Sum and Difference Frequencies Generation using Tunable Two- and Three-color Commensurate and Incommensurate Mid-infrared Pumps of Graphite Plasma

5.3.1. *Interaction of commensurate and incommensurate waves in the nonlinear media*

Historically, application of the two waves interacting in the nonlinear optical medium for improvement of high-order harmonic generation has initially been introduced during frequency conversion in the gas media. To amend HHG yield, the 800-nm-class lasers and second-order harmonics of this radiation were used at initial stages of those studies. Various schemes comprising generation of the 400 nm wave in a separate channel with further mixing with the 800 nm wave in the gases, as well as direct generation of second harmonic in thin BBO crystals followed with focusing of two beams in the converting medium were introduced (for example [13, 23, 25, 37, 54–56]). As underlined in [57], a strong harmonic generation in the case of two-color pump is possible due to the formation of a quasi-linear field, selection of a short quantum path component, which has a denser electron wave packet, and higher ionization rate compared with the single-color pump. The orthogonally polarized second field also participates in the modification of the trajectory of accelerated electron from being two-dimensional to three-dimensional that may lead to removal of the medium symmetry. With suitable control of the relative phase between the fundamental and second harmonic radiation, the latter field enhances the short path contribution while diminishing other electron paths, resulting in a clean spectrum of harmonics.

This approach has been applied for harmonic generation in laser-produced plasmas as well. The first observation of an increase of harmonic conversion efficiency in the narrow (∼0.3-mm-long) plasma plumes irradiated by an intense two-color femtosecond laser beam, wherein the fundamental field and its weak second harmonic (at the energy ratio of 50:1) were linearly polarized orthogonal to each other, was reported in [42]. Those and following studies [16,17] also showed the advantages of the TCP of plasmas with regard to the SCP. Note that most previous TCP experiments in gases and plasmas were restricted by application of solely Ti:sapphire lasers.

Next stage of these TCP studies could be related with application of the optical parametric amplifiers and their second harmonics. The application of second wave in those experiments can be even more useful compared with the 800 nm + H2 schemes since the mid-infrared pulses from OPA had the disadvantages of being less energetic and emitting at a longer wavelength region, which proved to be less efficient for HHG compared with the shorter-wavelength sources. Because of this the second-harmonic generation of the signal pulse from OPA could be successfully applied using the MIR + H2 TCP scheme for both gas and plasma HHG. Note that the majority of previous TCP schemes used the so-called commensurate waves, i.e. waves representing some integers with regard to each other.

Recently, application of incommensurate waves for TCP of gases was introduced and showed advantages both in harmonic yield and individual attosecond pulse generation from the multi-cycle sources [19, 58, 59]. The attractiveness of this approach was related with the availability of such multi-cycle sources in many laboratories and attempts to overpass the restrictions in generation of single attosecond pulses by the few-cycle radiation. In particular, it was found that the TCP using MIR and Ti:sapphire laser radiation greatly reduce the requirements for the lasers used for generation of shortest individual pulses through HHG [58, 60, 61].

Application of incommensurate two-color sources for HHG also has another advantage. It allows significantly increased number of sum and difference frequencies of interacting waves, which in turn may lead to the observation of resonance-induced enhancement of single component of those multi-harmonic sources. This process is mostly related with the ionic transitions of various materials possessing high oscillator strengths. From this point of view, the application of ablated materials with further use of the TCP comprising OPA and 800 nm pulses during propagation through the LPP may offer some advantages over MIR + H2 scheme, since this method allows the formation of conditions for above-mentioned resonance enhancement.

In this section, we analyze sum and difference harmonic generation using the mixing of tunable (1200–1600 nm) signal pulses from OPA and 810 nm ultrashort pulses in the extended laser-produced graphite plasma [62]. We show the optimization of high-order parametric and harmonic generation using various parameters (delay between two-color incommensurate pulses, ratio between pulses, use of parallel and orthogonal polarizations of interacting waves, etc.) and compare this scheme with the two-color pump of graphite plasma using commensurate waves (MIR and H2 pulses).

5.3.2. *Experimental results*

Experimental setup is described in previous section (Fig. 5.5). Both the signal and 810 nm pulses that propagated through the last amplifier of OPA were used for the TCP of LPP, together with weaker idler pulses, though the latter radiation in most cases insignificantly influenced the process of HHG. Each of these pumps was also separately used for comparison with two- and three-color pump schemes. The polarizations of signal and 810 nm pulses were orthogonal to each other. These beams were focused by the same 400-mm focal length lens into the extended graphite plasma. The difference in focal planes of two pumps (MIR and 810 nm) was ~3 mm due to dispersion of used lens and different divergences of pumps. Nevertheless, two focused beams sufficiently overlapped inside the plasma area assuming the confocal parameter of 810 nm radiation (12 mm) and plasma length (5 mm). The temporal overlap was also maintained at optimal conditions due to availability in variation of the delay of 810 nm pulse with regard to the parametric seeding pulse in the last stage of amplification of the OPA. Small group velocity dispersion of the focusing lens and input window in the MIR range allowed the delay between the pump pulses in the plasma area to be minimized.

Most of the experiments were carried out using the TCP and three-color pump of graphite LPP. The 0.5-mm-thick BBO crystal (type I) was installed inside the vacuum chamber on the path of focused signal pulse. The conversion efficiency of H2 pulses was ~20%. The spectral bandwidth of second-harmonic radiation was 22 nm. The overlapping of two pulses in plasma area was sufficient for the observation of the role of the second orthogonally polarized field influencing the whole process of HHG, due to small group velocity dispersion in the BBO crystal.

Upper panel of Fig. 5.12(a) shows the harmonic spectrum generated in the case of SCP (1310 nm, 1.2 mJ). The weak idler wave (2075 nm, 0.3 mJ) was also present in the plasma. However, its influence was insignificant due to smaller energy and above-mentioned wavelength rule, thus we did not see the harmonics or sum and difference frequencies attributed to this radiation at the used conditions. The addition of second field (810 nm, 3 mJ) led to significant enhancement of a whole yield of harmonics and appearance of the sum and difference frequency components (second panel). Dotted lines on this and other graphs show the harmonics

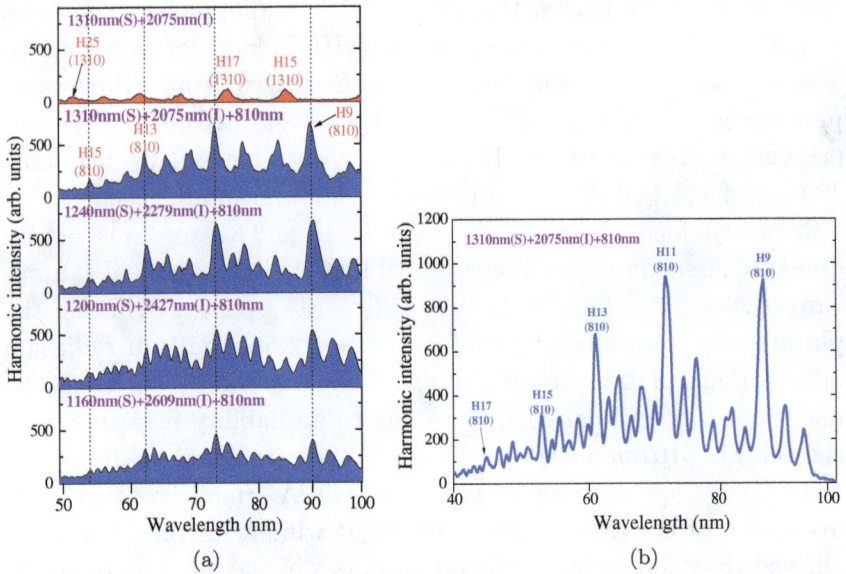

Fig. 5.12. (a) Harmonic spectra in the cases of (upper panel) generation from signal (1310 nm) pulse, (second panel) generation from mixture of 1310, 2075, and 810 nm pumps, and (three bottom panels) generation from 810 nm and tuned signal and idler pulses. The harmonic orders and corresponding sources are shown on the two upper panels. Dotted lines in this and other figures correspond to the odd harmonics of 810 nm wave. (b) Spectrum of harmonics generated from the mixture of 1310, 2075, and 810 nm pulses in the plasma area at variable overlap of strong 810 nm wave and weak signal and idler waves (see text). Reproduced from [62] with permission from Optical Society of America.

corresponded to the odd integers of 810 nm wave (from H9 to H15). One can see the appearance of two components between each two neighboring odd harmonics of 810 nm pump. In particular, the photon energies of two emission components between H9 ($\lambda = 90$ nm) and H11 ($\lambda = 73.6$ nm) of 810 nm pump almost exactly corresponded to $E_{H9} + E_{1310nm} = 14.73$ eV ($\lambda = 84.1$ nm) and $E_{H11} - E_{1310nm} = 15.88$ eV ($\lambda = 78.1$ nm). The same can be said about other groups of harmonic emission, contrary to the case of TCP using H2, which led to generation of $H_{integer}(1310$ nm$) \pm H2$ corresponding to the odd and even harmonics of MIR radiation. Thus in the present case we observed the parametric processes, which could be attributed to sum and difference frequency generation.

The energy difference between the 78.1 nm (15.88 eV) and 84.1 nm (14.73 eV) radiation generated between H9 and H11 did not exactly correspond to the energy of single photon of 1310 nm pulses (0.95 eV). This intolerance led to some additional frequency components, which can be recognized in the heterogeneous structure of the intermediate harmonics between the odd ones attributed to the stronger 810 nm field. Exact coincidence of parametric processes for sum and difference frequency components, which should show smooth spectral shapes of those emissions, could be expected in the case of the mixture of 1216 nm MIR pulses and 810 nm radiation. With these photons ($E_{1216nm} = 1.02$ eV), the difference of photon energies between four frequency components [H9$_{810nm}$(13.77 eV), H9 + MIR (14.79 eV), H11$_{810nm}$ − MIR (15.81 eV), and H11(16.83 eV)] should be equal to each other once we assume the parametric sum and difference processes of the involved waves. However, once we tuned the MIR radiation from 1310 nm towards the shorter wavelength region (∼1200 nm), a notable deviation from the above prediction was observed. At these conditions, we were able generating a set of multiple sum and difference waves. Three bottom panels of Fig. 5.12(a) show the spectra generated at the 1160, 1200, and 1240 nm pumps mixed with the 810 nm wave. Particularly, in the case of 1200 nm + 810 nm TCP (i.e. close to above predicted conditions), a few additional intermediate frequencies appeared between each odd harmonics of Ti:sapphire laser radiation. The energy difference between those components was equal to ∼0.5 eV, which approximately corresponded to the energy of the weak idler photon ($\lambda = 2427$ nm) present in the plasma alongside the signal and 810 nm radiation, which could be responsible for the appearance of those additional components, assuming a decrease of signal wave in this spectral region and non-optimal overlap of signal and 810 nm pulses.

This observation shows the involvement of three waves in the parametric generation of coherent XUV radiation. Note a significant difference in the energies of these waves, apart from the non-optimal overlap of these pulses in the plasma. Nevertheless, we achieved comparable sum, difference, and harmonic waves in the XUV [see two bottom panels in Fig. 5.12(a)]. An three-color pump significantly depends on the relative phases of interacting waves, delays between them, and spatial overlap. An example of the influence of relative phases and delays between interacting waves is shown in Fig. 5.12(b). The temporal deviation from the 'optimal' conditions corresponding to the efficient interaction of signal (1310 nm) and 810 nm waves caused by variable delay between these pulses led to the influence of another component (i.e. idler wave) and generation of additional frequencies between the odd harmonics of latter pump [compare this graph showing five components attributed to joint influence of signal, idler, and 810 nm waves and relatively smooth spectrum comprising two components attributed to generation from signal (1310 nm) and 810 nm waves, Fig. 5.12(a), second panel].

Once the delay between MIR and 810 nm pumps was minimized with regard to the signal component, as well as the focal position of pumps with regard to the plasma adjusted, the above predicted smooth and broadband harmonic spectrum appeared in the case of 1250 nm + 2246 nm + 810 nm pump [Fig. 5.13(a), upper panel; compare with other panels corresponding to the deviation from the 'optimal' relation between the interacting waves]. Note that in this case the influence of idler wave was not observed. We would also like to point out the equality of all harmonic components under these conditions in spite of significant difference in the energies of two pumps (4 mJ and 0.8 mJ in the case of 810 nm and 1250 nm pulses respectively) and above-mentioned wavelength-related rule of harmonic yield.

Two cases (1300 nm + 810 nm and 1300 nm + H2$_{\text{MIR}}$+810 nm) were compared as well to analyze the involvement of another third wave (H2$_{\text{MIR}}$) on the HHG in graphite plasma under similar experimental conditions. First case [Fig. 5.13(b), upper panel] showed a spectrum almost similar to the one described above (1310 nm + 810 nm pump; Fig. 5.12a, second panel). As already mentioned, it contained two frequency components between the odd harmonics of 810 nm wave. Then the BBO crystal was inserted in the path of pump waves under the conditions of the second harmonic phase matching for the MIR pulses ($\theta = 21°$). In that case additional harmonic

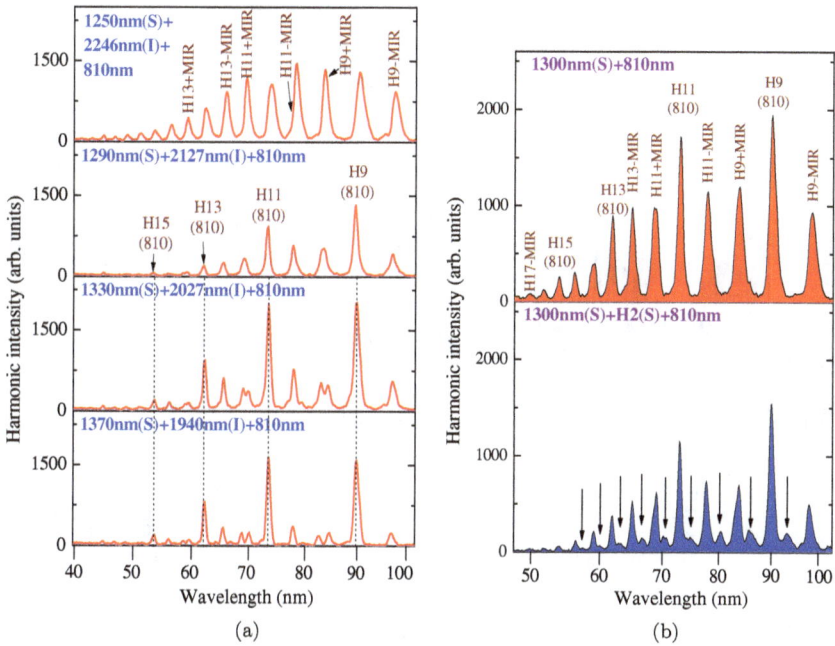

Fig. 5.13. (a) (upper panel) Harmonic spectrum using optimal relation between the wavelengths of two pumps (signal and 800 nm) and suitable delay between pulses. (Three bottom panels) "Non-optimal" conditions of HHG using TCP comprising different wavelengths of assistant field. (b) Harmonic spectra using two-color (1300 nm + 810 nm; upper panel) and three-color (1300 nm + 650 nm + 810 nm; bottom panel) pumps. Arrows in bottom panel show additional frequency components appeared after insertion of BBO in the path of 1300 and 810 nm pumps. The crystal was adjusted for the phase matching of H2 of the MIR pulses. Reproduced from [62] with permission from Optical Society of America.

components were observed between the harmonics shown in the upper panel [Fig. 5.13(b), bottom panel]. They were relatively weak compared with other frequency components due to non-optimized relation between the phases of three waves. This scheme is another type of three-color pump induced HHG. The harmonics lying exactly between odd integers of 810 nm pump may originate from the two-color pump (810 nm + 405 nm) since even at the orthogonally placed orientation of BBO crystal some small amount of second harmonic wave appeared during propagation of the strong 810 nm radiation through the crystal due to rotation of polarization.

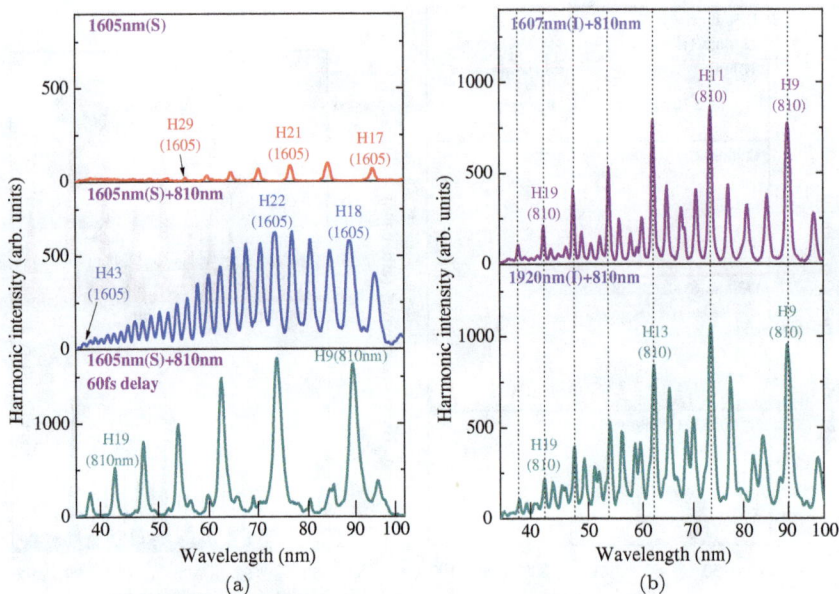

Fig. 5.14. (a) Harmonic spectra from 1605 nm signal pulses (upper panel) and orthogonally polarized TCP (1605 nm + 810 nm; middle panel). Bottom panel shows the spectrum using delayed 810 nm pulses with regard to seeded pulses of OPA, when the efficiency of parametric amplification significantly dropped. (b) Applications of idler pulses (1607 and 1920 nm) in TCP scheme (MIR + 810 nm) at parallel polarizations of two pumps. Reproduced from [62] with permission from Optical Society of America.

Another spectral region where the optimal conditions of two-color sum and difference frequency generation led to the smooth homogeneous and equidistant spectra is a degenerate regime of OPA, when the wavelengths of signal and idler waves became almost equal to each other ($\sim$1600 nm, which is in fact the doubled wavelength of the pump laser). Upper panel of Fig. 5.14(a) shows the harmonic spectrum in the case of 1605 nm signal pump. The application of orthogonally polarized two-color pump (MIR + 810 nm) led to generation of three almost equal frequency components between the odd ones originating from the 810 nm pump (second panel), in spite of significant difference between the energies of pump pulses (0.3 and 4 mJ correspondingly). Introducing a 60 fs delay of the 810 nm pulses with regard to the seeding radiation of OPA led to almost complete disappearance of parametric waves in the XUV spectra and generation of solely odd harmonics of 810 nm pump (third panel) due to weakened energy of the MIR signal radiation and worse overlap of the pulses.

During most of the above studies the orthogonally polarized pumps (signal waves from OPA and 810 nm pulses) were used. The application of parallel polarized pumps was analyzed using the TCP when the weak idler pulses were mixed with the 810 nm radiation in the graphite plasma. Figure 5.14(b) shows two spectra in the case of the 1607 and 1920 nm idler pulses mixed with the 810 nm pulses to pump the graphite plasma. The energies of these MIR pulses were notably weaker compared with signal waves, whilst the appearance of additional harmonic components in the case of TCP using these pulses was qualitatively similar to the application of stronger orthogonally polarized signal pulses in similar scheme.

The delay of 810 nm pump with regard to seeding pulses in the last amplifier of OPA was optimized to generate strongest signal pulses rather than idler ones. Under these conditions, the signal pulses were insufficiently overlapped with 810 nm pulses in the plasma area compared with the idler pulses due to polarization- and wavelength-dependent dispersion in the BBO amplifier of OPA, as well as the dispersion of other optics, such as focusing lens and input MgF_2 window. The temporal tuning of 810 nm pump allowed generation of sum and difference frequency components during mixing with either signal or idler pumps. Figure 5.15(a) shows the variation of harmonic spectra in the case of signal (1605 nm) + idler (1606 nm) + 810 nm pump for various temporal overlaps. At long delays (upper and bottom panels), only odd harmonics of 810 nm pump were generated. A decrease of delay from both sides led to appearance of sum and difference frequencies (second and fourth panels), until, at optimal (i.e. close to zero) delay it led to generation of almost equal harmonics (middle panel). One can see approximately similar spectra either from signal or idler pulses mixed with 810 nm pulses (second and fourth panels), due to equal wavelengths of those MIR pulses, which can be considered as the

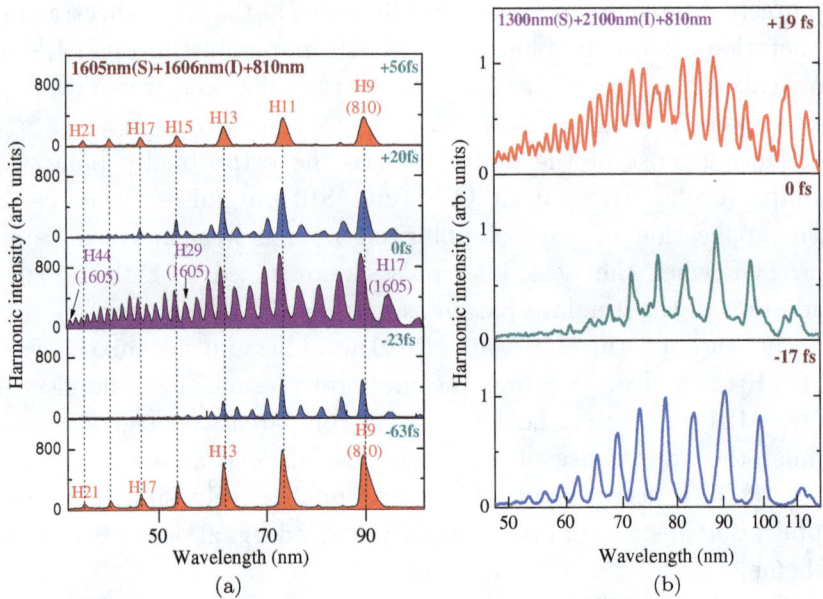

Fig. 5.15. (a) Harmonic spectra using 1605 nm (S) + 1606 nm (I) + 810 nm pump of plasma at different delays between 810 nm pulse and seeding pulses in OPA amplifier. (b) The same delay-dependent spectra in the case of 1300 nm (S) + 2100 nm (I) + 810 nm pump. Reproduced from [62] with permission from Optical Society of America.

commensurate waves with regard to the Ti:sapphire laser radiation. Another pattern appeared in the case of incommensurate waves. In particular, similar variations of delay between 810 nm and 1300 nm + 2100 nm pumps led to involvement of either former or latter MIR pulses in parametric interaction with the 810 nm pump. Upper panel of Fig. 5.15(b) shows the harmonic spectrum generated during better overlap of the 810 nm and idler pulses compared with the signal ones. Those harmonics were identified as odd and even integers of 2100 nm pump, as well as sum and difference components. Once the delay of 810 nm pump was optimized with regard to signal pulses, then smooth pattern of harmonics originating from interaction of those waves in plasma appeared again (bottom panel), similarly to Fig. 5.13(b) (upper panel). At intermediate ('zero') delay, HHG spectrum comprised the frequency components attributed to the mixture of above processes (middle panel), with preferable influence of stronger (signal) component.

5.3.3. *Discussion*

Summarizing the analysis of those studies of plasma harmonics using various combinations of two- and three-color pumps using MIR and 810 nm pulses, we would like to underline the advantages and disadvantages of this method of multiple waves interaction compared with the conventional one, which used two commensurate waves (i.e. fundamental and second-harmonic pulses). By comparing these two schemes one has to admit a difference of the relative intensities of the second additional wave with regard to the original one. In the case of TCP using second-harmonic generation followed with the sum frequency generation the ratio of second-harmonic and fundamental waves of commonly used Ti:sapphire lasers was of the order of 0.05 to 0.2. This ratio in the case of MIR pulses and their second-harmonic emission was larger, but maintained at approximately same level (0.1–0.3). As already mentioned, the weak second-harmonic photons were considered as the assistant field, which is sufficient to improve the ionization rate of emitters, as well as to amend the recombination cross-section of accelerated electrons [57]. In the case described in present section, the second wave (810 nm) was notably stronger than the signal waves from OPA. The 810 nm/MIR ratio was ∼10 to 3, thus one can rather consider MIR pulses as the assistant field to the strong 810 nm pulses [59]. So what can one practically gain in that case?

Figure 5.16(a) summarizes the comparative features of different approaches of the SCP and TCP of LPP for the HHG. Two upper panels show the harmonic spectra using SCP (810 nm and 1310 nm respectively). Obviously, in the former case one can see larger harmonic yield due to stronger 810 nm pulses, as well as due to above-mentioned $I_H \propto \lambda^{-5}$ rule. So we do not compare these two cases anymore. The use of TCP by introducing BBO on the path of 1310 nm pump led to significant enhancement of harmonic yield (third panel) compared with the SCP (second panel). Note that the

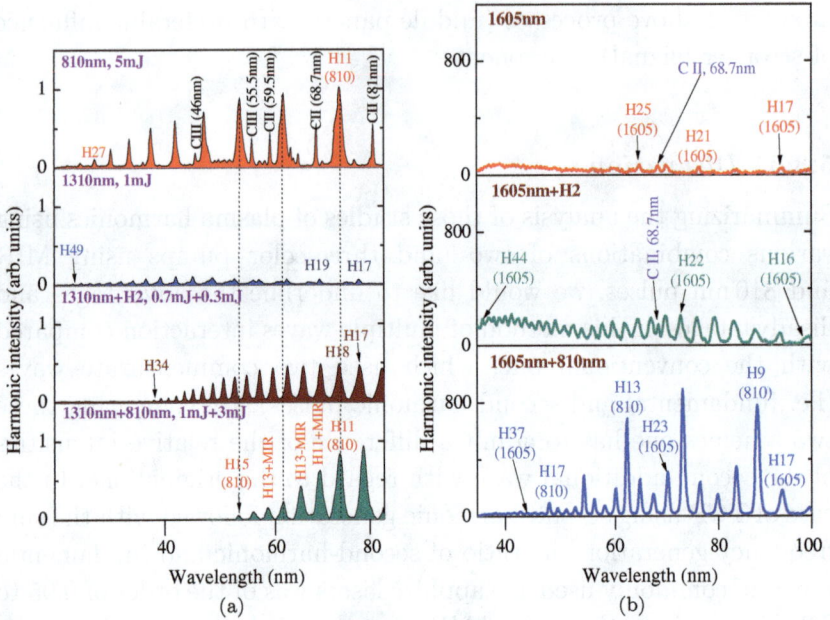

Fig. 5.16. (a) Summary of harmonic spectra studies using 1310 nm pump. For comparison, upper panel shows the HHG spectrum obtained using 810 nm, 5 mJ pulses. Second, third and fourth panels show the harmonic spectra using SCP (1310 nm, 1 mJ), TCP (1310 nm + 655 nm, 0.7 mJ + 0.3 mJ), and TCP (1310 nm + 810 nm, 1 mJ + 3 mJ). Spectral emission lines from graphite plasma are shown in the upper panel. (b) Summary of harmonic spectra studies using 1605 nm pump. Upper panel shows very weak harmonics from SCP (1605 nm). Second and third panels show TCP of plasma using MIR + H2 and MIR + 810 nm schemes. Reproduced from [62] with permission from Optical Society of America.

difference in harmonic yields from SCP and TCP became larger once we started to use the metal targets, which could be explained by the involvement of different emitters of harmonics in the cases of metal- and graphite-ablated plasmas [9–11, 63]. Bottom panel shows the HHG spectrum using MIR + 810 nm configuration. The overall harmonic yield in that case was larger compared with MIR + H2 induced HHG. The MIR + 810 nm pump using incommensurate waves allowed generation of various high-order sum and difference components of the interacting waves.

Similar conclusions could be drawn in the case when MIR pulses become 'commensurate' compared with the second field [∼1600 nm and ∼800 nm; Fig. 5.16(b)]. The measurements under similar conditions showed that, while in the case of SCP (1600 nm) extremely weak harmonics could be achieved (upper panel), the insertion of BBO in the path of this beam notably enhances the HHG yield, as well as harmonic cutoff, and allows generation of equal odd and even orders (middle panel). The application of proposed TCP approach (1600 nm + 810 nm) in plasma harmonic studies further enhanced the harmonic yield of the MIR-related components of harmonic spectra, alongside the strong odd harmonics from the 810 nm wave (bottom panel). The tuning of MIR pulses was advantageous for comparison of the pairs of commensurate and incommensurate pumps for plasma HHG, as well as allowed the application of either orthogonal or parallel polarizations of two waves using simplified experimental arrangements.

The comparative studies of our TCP scheme with the conventional one have shown both advantages [stronger yield of the MIR-related components of generated XUV spectrum, absence of additional nonlinear medium (BBO) for formation of second field, much denser emission spectrum] and disadvantages (less harmonic cutoff) of the MIR + 810 nm pump compared with the MIR + H2 scheme. The former scheme based on two-color laser field synthesis [58] has previously been proven to be suitable for nonlinear pump/probe experiments in gases that demonstrated the capability of time-resolved attosecond dynamic processes using individual attosecond pulses generated in gaseous media [19]. The implementation of similar technique for plasma harmonics may reveal new interesting features including the peculiar properties of individual emitters thus paving the way for laser ablation induced time-resolved HHG spectroscopy. Plasma harmonic approach allows thousands of solids to be used as the targets for ablation, contrary to gas harmonic approach, thus creating the conditions for material science studies.

The proposed TCP of plasmas significantly enhances the density of harmonics per wavelength unit in the XUV and increases the harmonic yield. A useful application of this technique could be generation of attosecond pulses using multi-cycle radiation. The experimental conditions almost coincided with the results calculated for in-gas harmonics [59] using the 1330 nm + 800 nm and 1600 nm + 800 nm TCP, which predicted single attosecond pulse generation under these conditions.

The dense emission spectrum shown in the case of 1160 nm + 810 nm pump [Fig. 5.12(a), bottom panel] is a combination of high-order difference and sum frequency generation, similar to those reported in the case of in-gas HHG studies [58]. The resolution of the used XUV spectrometer was good enough to fully resolve these frequency components. The appearance of such a dense spectrum of coherent emission can be advantageous for the formation of the conditions for continuum generation leading to single attosecond pulses using multi-cycle pumps at optimum carrier-envelope phase of driving pulses and suitable relative phase of two pumps.

5.4. Conclusions to Chapter 5

New relevant knowledge of these studies is related with the application of suitable ablating material using the optimized experimental conditions of plasma formation, which allowed generation of stable high-order harmonics in the plasma produced on the surface of super-hard material. This knowledge is applicable for the practical use of the HHG technique, as well as for the fundamental understanding of harmonic generation in the molecular plasmas.

The harmonic generation up to the 27th order ($\lambda = 29.9$ nm) of 806 nm driving pulses in the boron carbide LPP was demonstrated. The advantages and disadvantages of this target compared with the ingredients comprising B_4C (solid boron and carbon) have been analyzed by studying the plasma emission and harmonic spectra from three plasma species. The comparison of these plasma formations, from the point of view of the stability of harmonic emission, showed that boron carbide plasma is the best choice among three ablated targets. Different schemes of the commensurate and incommensurate two-color pumps of B_4C plasma were compared, particularly using the second harmonics of 806 nm laser and optical parametric amplifier as the assistant fields, as well as the sum and difference frequency generation demonstrated using the mixture of two laser sources. It was also shown that the coincidence of the harmonic and plasma emission wavelengths during ablation of boron carbide does not lead to enhancement or decrease of the conversion efficiency of this harmonic. The spatial characterization of harmonics has shown their modification depending on the conditions of HHG. This allowed the definition of the impeding processes, which influence the HHG conversion efficiency.

We also discussed HHG in laser-produced silver, gold, and zinc plasma plumes using 1310 nm emission from optical parametric amplifier and its second harmonic. This approach allowed us to increase significantly the high-order harmonic yield in comparison with the single-color pump. The highest increase of more than an order of magnitude is found for silver plasma; the enhancement is higher for higher-order harmonics. These experimental tendencies, as well as the gain values, are reasonably well reproduced in the calculations. The agreement was achieved due to the development of the HHG theory both in microscopic and macroscopic aspects.

It was found that the microscopic HHG spectrum calculated using the widely used assumption that the ground state of the generating particle is the $1s$ state dramatically differs from the experimental one. In contrast to that, the shape of the spectrum obtained with the actual quantum numbers of the outer electron is quite close to that of the experimental one. Thus, proper description of the ground state is crucial for the correct reproduction of the HHG spectrum in theoretical calculations.

The pronounced dependence of the microscopic response on the relative phase between the frequency components of a two-color field strongly influences the properties of the macroscopic signal because the relative phase changes during the propagation. Due to this dependence, only a limited range of the relative phases contributes to the phase-matching integral (5.13), preventing the vanishing of the harmonic macroscopic signal after propagation over the doubled coherence length (calculated for this harmonic in the single-color field).

These theoretical studies of the microscopic response clarify the important aspect of the HHG enhancement in the two-color orthogonally polarized fields with comparable intensities of the components. In a single-color field, half of the electronic trajectories never come back to the origin and the majority of the other trajectories return with low velocities. As a result, only a small fraction of electrons are responsible for HHG. In contrast, in the two-color case (with properly chosen relative phase) the total field maximum is shifted so that all the electrons released near the field maximum return with high energies. Though deflected in the orthogonal direction by the second field, the returning electron wave packets are large enough to recombine nearly as efficiently as in the single-color case. Thus, enhanced population of the trajectories recombining with high energy in conjunction with comparable recombination efficiency leads to the increase of the HHG yield in a two-color field.

Finally, we have demonstrated high-order sum and difference frequencies generation using the mixtures of tunable (1200–1600 nm) pulses from optical parametric amplifier and 810 nm ultrashort pulses in the extended laser-produced graphite plasma using the two- and three-color pumps. We have analyzed the optimization of high-order harmonic generation using various parameters (delay between two-color incommensurate and commensurate waves, ratio between pulse intensities, use of parallel and orthogonal polarizations of the interacting pulses, etc.) and comparison with commonly used two-color pump scheme (mid-infrared and second-harmonic pulses). We have shown that, even regardless of non-optimal spatio-temporal overlap of two or three sources of radiation in the plasmas, the application of the proposed technique allows a significant broadening of the number of different combinations of interacting waves and generation of sum and difference frequency harmonics in the extreme ultraviolet region. Notice that recent studies [64] have analyzed various laser-produced metal plasmas (palladium, zinc, lead, silver, and cadmium) for high-order sum and difference harmonic generation using the mixing of tunable mid-infrared signal and idler pulses and 810 nm ultrashort pulses. Those studies confirmed the above conclusions about the perspectives of the used approach for the growth of harmonic conversion efficiency and broadening of harmonic spectra.

References

[1] M. Oujja, A. Benítez-Canete, M. Sanz, I. Lopez-Quintas, M. Martín, R. de Nalda, and M. Castillejo, *Appl. Surf. Sci.* **336**, 53 (2015).

[2] M. Oujja, R. de Nalda, M. Lopez-Arias, R. Torres, J. P. Marangos, and M. Castillejo, *Phys. Rev. A* **81**, 043841 (2010).

[3] R. de Nalda, M. Lopez-Arias, M. Sanz, M. Oujja, and M. Castillejo, *Phys. Chem. Chem. Phys.* **13**, 10755 (2011).

[4] I. Lopez-Quintas, M. Oujja, M. Sanz, M. Martin, R. A. Ganeev, and M. Castillejo, *Appl. Surf. Sci.* **278**, 33 (2013).

[5] R. A. Ganeev, P. A. Naik, H. Singhal, J. A. Chakera, M. Kumar, U. Chakravarty, and P. D. Gupta, *Opt. Commun.* **285**, 2934 (2012).

[6] R. A. Ganeev, G. S. Boltaev, N. K. Satlikov, I. A. Kulagin, R. I. Tugushev, and T. Usmanov, *J. Opt. Soc. Am. B* **29**, 3286 (2012).

[7] W. Theobald, C. Wülker, F. R. Schäfer, and B. N. Chichkov, *Opt. Commun.* **120**, 177 (1995).

[8] R. A. Ganeev, M. Suzuki, and H. Kuroda, *Opt. Commun.* **370**, 6 (2016).

[9] L. B. Elouga Bom, Y. Pertot, V. R. Bhardwaj, and T. Ozaki, *Opt. Express* **19**, 3077 (2011).

[10] Y. Pertot, L. B. Elouga Bom, V. R. Bhardwaj, and T. Ozaki, *Appl. Phys. Lett.* **98**, 101104 (2011).

[11] R. A. Ganeev, T. Witting, C. Hutchison, F. Frank, P. V. Redkin, W. A. Okell, D. Y. Lei, T. Roschuk, S. A. Maier, J. P. Marangos, and J. W. G. Tisch, *Phys. Rev. A* **85**, 015807 (2012).

[12] R. A. Ganeev, M. Suzuki, S. Yoneya, and H. Kuroda, *Appl. Phys. Lett.* **105**, 041111 (2014).

[13] I. J. Kim, C. M. Kim, H. T. Kim, G. H. Lee, Y. S. Lee, J. Y. Park, D. J. Cho, and C. H. Nam, *Phys. Rev. Lett.* **94**, 243901 (2005).

[14] R. A. Ganeev, H. Singhal, P. A. Naik, J. A. Chakera, H. S. Vora, R. A. Khan, and P. D. Gupta, *Phys. Rev. A* **82**, 053831 (2010).

[15] M. Negro, C. Vozzi, K. Kovacs, C. Altucci, R. Velotta, F. Frassetto, L. Poletto, P. Villoresi, S. De Silvestri, and V. Tosa, *Laser Phys. Lett.* **8**, 875 (2011).

[16] R. A. Ganeev, C. Hutchison, A. Zaïr, T. Witting, F. Frank, W. A. Okell, J. W. G. Tisch, and J. P. Marangos, *Opt. Express* **20**, 90 (2012).

[17] R. A. Ganeev, V. V. Strelkov, C. Hutchison, A. Zaïr, D. Kilbane, M. A. Khokhlova, and J. P. Marangos, *Phys. Rev. A* **85**, 023832 (2012).

[18] R. A. Ganeev, M. Suzuki, and H. Kuroda, *J. Phys. B: At. Mol. Opt. Phys.* **47**, 105401 (2014).

[19] E. Takahashi, Pengfei Lan, O. D. Mücke, Y. Nabekawa, and K. Midorikawa, *Nature Commun.* **4**, 2691 (2013).

[20] T. T. Liu, T. Kanai, T. Sekikawa, and S. Watanabe, *Phys. Rev. A* **73**, 063823 (2006).

[21] N. Dudovich, J. L. Tate, Y. Mairesse, D. M. Villeneuve, P. B. Corkum, and M. B. Gaarde, *Phys. Rev. A* **80**, 011806 (2009).

[22] X. He, J. M. Dahlström, R. Rakowski, C. M. Heyl, A. Persson, J. Mauritsson, and A. L'Huillier, *Phys. Rev. A* **82**, 033410 (2010).

[23] J. Mauritsson, P. Johnsson, E. Gustafsson, A. L'Huillier, K. J. Schafer, and M. B. Gaarde, *Phys. Rev. Lett.* **97**, 013001 (2006).

[24] C. M. Kim, I. J. Kim, and C. H. Nam, *Phys. Rev. A* **72**, 033817 (2005).

[25] D. Charalambidis, P. Tzallas, E. P. Benis, E. Skantzakis, G. Maravelias, L. A. A. Nikolopoulos, A. P. Conde, and G. D. Tsakiris, *New J. Phys.* **10**, 025018 (2008).

[26] L. Brugnera, D. J. Hoffmann, T. Siegel, F. Frank, A. Zaïr, J. W. G. Tisch, and J. P. Marangos, *Phys. Rev. Lett.* **107**, 153902 (2011).

[27] L. Brugnera, F. Frank, D. J. Hoffmann, R. Torres, T. Siegel, J. G. Underwood, E. Springate, C. Froud, E. I. C. Turcu, J. W. G. Tisch, and J. P. Marangos, *Opt. Lett.* **35**, 3994 (2010).

[28] H. Eichmann, A. Egbert, S. Nolte, C. Momma, B. Wellegehausen, W. Becker, S. Long, and J. K. McIver, *Phys. Rev. A* **51**, R3414 (1995).

[29] S. Long, W. Becker, and J. K. McIver, *Phys. Rev. A* **52**, 2262 (1995).

[30] W. Becker, B. N. Chichkov, and B. Wellegehausen, *Phys. Rev. A* **60**, 1721 (1999).

[31] D. B. Milošević, W. Becker, and R. Kopold, *Phys. Rev. A* **61**, 063403 (2000).

[32] Z. Zeng, Y. Cheng, X. Song, R. Li, and Z. Xu, *Phys. Rev. Lett.* **98**, 203901 (2007).

[33] M. Murakami, O. Korobkin, and M. Horbatsch, *Phys. Rev. A* **88**, 063419 (2013).

[34] D. J. Hoffmann, C. Hutchison, A. Zaïr, and J. P. Marangos, *Phys. Rev. A* **89**, 023423 (2014).

[35] J. Henkel and M. Lein, *Phys. Rev. A* **92**, 013422 (2015).

[36] J.-W. Geng, W.-H. Xiong, X.-R. Xiao, L.-Y. Peng, and Q. Gong, *Phys. Rev. Lett.* **115**, 193001 (2015).

[37] T. Pfeifer, L. Gallmann, M. J. Abel, D. M. Neumark, and S. R. Leone, *Opt. Lett.* **31**, 975 (2006).

[38] P. V. Ignatovich, V. T. Platonenko, and V. V. Strelkov, *Laser Phys.* **9**, 570 (1999).

[39] K. Schiessl, E. Persson, A. Scrinzi, and J. Burgdörfer, *Phys. Rev. A* **74**, 053412 (2006).

[40] C. Jin, G. J. Stein, K.-H. Hong, and C. D. Lin, *Phys. Rev. Lett.* **115**, 043901 (2015).

[41] B. Schutte, P. Weber, K. Kovacs, E. Balogh, B. Major, V. Tosa, S. Han, M. J. J. Vrakking, K. Varju, and A. Rouzee, *Opt. Express* **23**, 33947 (2015).

[42] R. A. Ganeev, H. Singhal, P. A. Naik, I. A. Kulagin, P. V. Redkin, J. A. Chakera, M. Tayyab, R. A. Khan, and P. D. Gupta, *Phys. Rev. A* **80**, 033845 (2009).

[43] R. A. Ganeev, M. Suzuki, and H. Kuroda, *Eur. Phys. J. D* **70**, 21 (2016).

[44] A. S. Emelina, M. Y. Emelin, R. A. Ganeev, M. Suzuki, H. Kuroda, and V. V. Strelkov, *Opt. Express* **24**, 13971 (2016).

[45] M. Rubenchik, M. D. Feit, M. D. Perry, and J. T. Larsen, *Appl. Surf. Sci.* **127**, 193 (1998).

[46] J. Tate, T. Auguste, H. G. Muller, P. Salières, P. Agostini, and L. F. DiMauro, *Phys. Rev. Lett.* **98**, 013901 (2007).

[47] P. Lan, E. J. Takahashi, and K. Midorikawa, *Phys. Rev. A* **81**, 061802 (2010).

[48] R. A. Ganeev, M. Baba, M. Suzuki, and H. Kuroda, *Phys. Lett. A* **339**, 103 (2005).

[49] A. S. Emelina, M. Y. Emelin, and M. Y. Ryabikin, *Quantum Electron.* **44**, 470 (2014).

[50] S. Emelina, M. Y. Emelin, and M. Y. Ryabikin, *J. Opt. Soc. Am. B* **32**, 2478 (2015).

[51] M. Lewenstein, P. Balcou, M. Y. Ivanov, A. L'Huillier, and P. B. Corkum, *Phys. Rev. A* **49**, 2117 (1994).

[52] V. V. Strelkov, M. A. Khokhlova, A. A. Gonoskov, I. A. Gonoskov, and M. Y. Ryabikin, *Phys. Rev. A* **86**, 013404 (2012).

[53] V. V. Strelkov, *Phys. Rev. A* **74**, 013405 (2006).

[54] E. Cormier and M. Lewenstein, *Eur. Phys. J. D* **12**, 227 (2000).

[55] Y. Yu, X. Song, Y. Fu, R. Li, Y. Cheng and Z. Xu, *Opt. Express* **16**, 686 (2008).

[56] X.-S. Liu and N.-N. Li, *J. Phys. B: At. Mol. Opt. Phys.* **41**, 015602 (2008).

[57] I. J. Kim, G. H. Lee, S. B. Park, Y. S. Lee, T. K. Kim, C. H. Nam, T. Mocek and K. Jakubczak, *Appl. Phys. Lett.* **92**, 021125 (2008).

[58] E. Takahashi, P. Lan, O. D. Mücke, Y. Nabekawa and K. Midorikawa, *Phys. Rev. Lett.* **104**, 233901 (2010).

[59] P. Lan, E. Takahashi, and K. Midorikawa, *Phys. Rev. A* **82**, 053413 (2010).

[60] T. Pfeifer, L. Gallmann, M. J. Abel, P. M. Nagel, D. M. Neumark and S. R. Leone, *Phys. Rev. Lett.* **97**, 163901 (2006).

[61] B. Kim, J. Ahn, Y. Yu, Y. Cheng, Z. Zhizhan and D. E. Kim, *Opt. Express* **16**, 10331 (2008).

[62] R. A. Ganeev, *J. Opt. Soc. Am. B* **33**, E93 (2016).

[63] M. A. Fareed, S. Mondal, Y. Pertot and T. Ozaki, *J. Phys. B: At. Mol. Opt. Phys.* **49**, 035604 (2016).

[64] R. A. Ganeev, *Opt. Spectrosc.* **120**, 766 (2016).

Summary: Perspectives

It becomes obvious from the description of the studies of HHG using propagation of MIR pulses through various plasmas that the developments in this direction are aimed for the improvement of harmonic yield, precise study of resonance effects during fine tuning of driving pulses to the autoionizing resonances, quasi-phase-matching of MIR and XUV pulses, etc. Various evidences presented in this book show the advanced features of the proposed approach of application of the longer-wavelength pulses for the analysis of various processes of "MIR-plasma" interaction. To not repeat again the conclusions already emphasized at the end of each chapter, we would like to point out the issue which has yet to find sufficient attention — the availability of "MIR-plasma" interaction for attosecond pulses generation. Probably, it is one of the most intriguing topics which could be the next step for plasma harmonic studies.

Attosecond plasma harmonics spectroscopy using MIR pulses could be aimed at developing a new method for material studies based on high-order harmonic generation in laser-induced plasma plumes. This XUV harmonic radiation has been shown to contain information on the electronic structure of the generating medium. The scientific originality of this approach is related with the novel way in ultrashort pulses generation using the harmonic generation in laser-produced plasmas, as well as application of this approach for the attosecond spectroscopy of ablated materials.

There are a few main aspects which distinguish the proposed approach with regard to the already reported studies of plasma harmonics and show the competiveness of plasma harmonic concept over gas harmonic (i.e. resonance enhancement, application of clustered media, quasi-phase-matching of specifically modulated plasma plumes, etc.) for generation of ultrashort pulses in the XUV region. The plasma harmonic approach has been exploited for producing an efficient source of short-wavelength ultra-short pulses for various applications. Early studies concerning HHG in a plasma medium found that the best conditions for harmonic generation were achieved in low-excited and weakly ionized plasma, because the limiting processes governing the dynamics of laser wavelength conversion play a minor role in that case. The orders of harmonics obtained so far in plasma media range into the sixties and seventies (10–13 nm) using Ti:sapphire lasers. The highest harmonic order obtained so far has a wavelength of 7.5 nm and has been demonstrated in manganese plasma. The typical HHG conversion efficiency in the plateau region amounts to 10^{-5}. In the case of resonantly enhanced harmonic generation the conversion efficiency for an individual harmonic can reach 10^{-4}. To achieve higher harmonic conversion efficiency, the length of the medium can be increased, provided that the phase mismatch between the laser field and the harmonic radiation remains low. For the short-sized media, where the reabsorption can be neglected, and for the optimum phase-matching conditions, the harmonic intensity increases as the square of the medium length. However, once the medium length exceeds the coherence length, the harmonic intensity starts to oscillate due to variations of phase mismatch. To analyze this process in the laser-produced plasmas for attosecond pulses generation, one has to carefully define the best conditions of plasma HHG in the extended medium, while taking into account the peculiar properties of different materials used for laser ablation. The application of spatially modulated heating pulses, or perforated targets, could be an attractive method for the multiple plasma jets formation for the quasi-phase-matching conditions. Investigations in this field of plasma harmonics are making rapid strides and opening exciting possibilities in the nearest future. Plasma harmonics became an important part of the studies carried out in various laboratories worldwide and the achievements of present-day plasma harmonic studies motivate for further developments of this technique.

High-order harmonic generation occurs when an intense short laser pulse is focused into a gas of atoms. It was first shown theoretically and then experimentally that the generated radiation consists of pulses with attosecond duration. This attosecond nature of the harmonic emission, along with its short wavelength, excellent coherence and high brilliance provide a strong motivation for applications, some of which have already been demonstrated.

Another medium for harmonic generation and formation of ultrashort pulses is the laser-produced plasma. The first measurement of the attosecond emission generated from underdense plasma produced on a solid target was reported in 2011. Those studies have demonstrated generation of high-order harmonics of a femtosecond Ti:sapphire laser focused in a weakly ionized underdense chromium plasma. It was shown using the RABITT technique that the 11th to the 19th harmonic orders form in the time domain an attosecond pulse train with each pulse having 300 as duration, which is only 1.05 times the theoretical Fourier transform limit. It has been suggested that, besides its fundamental interest, high-order harmonic generation in plasma plumes could thus provide an intense source of attosecond pulses for applications. The first temporal characterization of the attosecond emission from a tin plasma under near-resonant conditions for two different resonance detunings has shown that the resonance considerably changes the relative phase of neighboring harmonics.

The plasma harmonic technique allows application of a variety of numerous targets for HHG, including some with very strong radiative transitions. It has been underlined that only the spectral intensity of high-order harmonics from resonant plasmas has been experimentally characterized and the important question of whether they offer a route to intense XUV femtosecond or attosecond pulses still remains unanswered. Thus the optimization of the MIR driving wave with regard to the resonances possessing strong oscillator strength could be a route to the formation of attosecond pulses in laser–plasma interactions.

The so-called 'self-probing' schemes to extract structural and dynamic information about the generating system from intensity, phase and polarization measurements of high harmonics have already been reported in the case of gas harmonic experiments. The implementation of this concept for the case of plasma harmonics in the case of MIR pump widely broadens the subjects of study due to overwhelming prevalence of the ablated solid species used in the latter case over a few gases. The theoretical aspects of attosecond pulses generation in plasmas are similar to the gas harmonic concept. Regarding the plasma harmonic concept itself, the three-step model fully describes the principles of HHG in this medium. Multiple theoretical descriptions of plasma harmonics and some processes in the plasmas during HHG were reported in numerous papers.

The key enabling technology for this research is the capability of laser ablation to put atoms and molecules from solids in gas phase at densities sufficient for HHG attosecond spectroscopy measurements. In particular the future projects could be articulated in the following steps: (a) new approaches in MIR-induced HHG from plasma plumes, (b) development of new schemes for resonantly enhanced harmonics using MIR pulses, (c) comparison between MIR-induced HHG in standard gas jets and plasma plumes, (d) study of the plasma formation dynamics by HHG spectroscopy, (e) harmonic generation from laser plumes containing nanoparticles, (f) characterization of the MIR-induced XUV ultrashort pulses generated in the plasma plumes during optimized HHG, (g) application of attosecond pulses for nonlinear spectroscopy of ablated materials. These studies may include the use of various techniques developed earlier in the case of gas harmonics (polarization gating, circular polarization induced HHG, molecular orientation related modulation of the phase of harmonics, etc.). The advantage of plasma harmonic approach over gas harmonic one is related with the dramatically larger amount of available solid species (over a few gases commonly used in the latter case), which widely broaden the area of investigations of the materials.

The ultrashort pulses have attracted strong interest, since they can record frozen snapshots of ultrafast electron dynamics, which have already been applied to study photoionization of atoms and molecules with attosecond resolution. The generation of an intense, isolated attosecond pulse is of importance for attosecond nonlinear, attosecond pump and attosecond probe experiments.

It has already been demonstrated that a continuum of high-order harmonics generated in the laser-produced plasmas could be generated with an 8 fs Ti:sapphire laser. Such continuum is a signature of isolated attosecond pulse. Though a continuum spectrum is not a direct demonstration of the generation of a single attosecond pulse, it is the first step towards the generation of intense isolated attosecond pulses, by combining the plasma harmonic technique and the double optical gating technique. Currently, the major problem is related to the limited duration for a stable continuum generation due to the manual translation of the solid target surface, since harmonic stability is not long enough for single attosecond pulse characterization or the carrier envelope phase scans. In this connection notice that the stability of plasma harmonics has been significantly improved by using the technique of ablating the rotating targets, which allowed stable harmonic generation lasting over 1 million pulses. Thus this method offers an opportunity in the analysis of the temporal structure of plasma harmonics.

Recently, application of incommensurate waves for two-color pump of gases was introduced and showed advantages both in harmonic yield and individual attosecond pulses generation from multi-cycle sources. The attractiveness of this approach is related with the availability of such multi-cycle sources in many laboratories and attempts to overpass the restrictions in generation of single attosecond pulses by the few-cycle radiation. In particular, it was found that the two-color pump using MIR and Ti:sapphire laser radiation greatly reduce the requirements for the lasers used for generation of shortest individual pulses through HHG.

Application of incommensurate two-color sources for plasma HHG also has another advantage. It allows the number of sum and difference frequencies of interacting waves to be significantly increased, which in turn may lead to the observation of resonance-induced enhancement of single component of those multi-harmonic

sources. This process is mostly related with the ionic transitions
of various materials possessing high oscillator strengths. From this
point of view, the application of ablated materials with further use
of the two- or three-color pump comprising OPA and 800 nm pulses
during propagation through the plasma may offer some advantages
over MIR + H2 scheme, since this method allows the formation of
perfect conditions for above-mentioned resonance enhancement of
harmonics. Notice that the MIR + 810 nm pump using incommensurate waves allowed the generation of various high-order sum and
difference components of the interacting waves using two- and three-
color pumps of plasmas, which significantly enhances the density
of harmonics per wavelength unit in the XUV and increases the
harmonic yield. A useful application of this technique could be
generation of attosecond pulses using multi-cycle radiation.

The advantages gained during current stages of these studies
will be extremely useful for implementation of this technique for
attosecond pulses generation and plasma spectroscopy. The novelty
of the majority of proposed ideas and availability of resources
(equipment, manpower, experience in this field) allow us to expect
a successful implementation of attosecond plasma harmonic spectroscopy in various laboratories worldwide. The proposed plasma
harmonic spectroscopy allows some important impulse to be added
to those studies, by using the peculiarities of *in-plasma HHG*.

The future studies on these topics will be instrumental to
establish the new field of HHG spectroscopy in laser plasmas in the
XUV spectral range. Plasma HHG will allow an analysis of a wide
range of molecules, from simple diatomic molecules to more complex
samples, such as aromatic molecules and small peptides. Previous
studies of plasma harmonics using nucleobases of DNA (uracil,
thymine) have confirmed the survival of complex molecules after
appropriately maintained laser-induced ablation. This allows us to
expect that some other fragile species can be exploited using similar
techniques. Whilst the advantages of plasma harmonic spectroscopy
for material science are obvious and are described throughout the
book, potential results of these studies may include applications
in plasma diagnostics, time-resolved measurements, biology and

medicine, microscopy, and other competitive areas. The potential applications of efficient XUV coherent sources and the development of plasma harmonic spectroscopy may lead to interesting perspectives in nanofabrication and surface diagnostics, as well as in further development of attosecond sources.

Index